ELEMENTARY ALGEBRA

A First Course in the Algebra of Real Numbers

ELEMENTARY ALGEBRA

A First Course in the Algebra of Real Numbers

PAUL J. MEERT
Bakersfield College

DICKENSON PUBLISHING COMPANY, INC.
Encino, California and Belmont, California

ISBN-0-8221-0126-2

Library of Congress Catalog Card Number: 73-92478

Printed in the United States of America

Printing (last digit): 9 8 7 6 5 4 3 2 1

To My Wife

Contents

Preface   *xiii*

Chapter 1: The Structure of Mathematics   *1*

1-1   Mathematical systems   *1*
1-2   Undefined terms   *2*
1-3   The set of real numbers   *2*
1-4   Sets   *5*
1-5   Some axioms of the real number system   *6*

Chapter 2: Four Operations on the Real Numbers   *10*

2-1   Representing real numbers with variables   *10*
2-2   Sums and terms   *11*
2-3   Factors and exponents   *11*
2-4   The absolute value of a real number   *13*
2-5   Addition of real numbers   *14*
2-6   Subtraction of real numbers   *16*
2-7   Multiplication of real numbers   *19*
2-8   Division of real numbers   *20*
2-9   Order of performing operations   *21*
2-10   Evaluation of algebraic expressions   *23*
          Cumulative review: chapters 1 and 2   *24*

Chapter 3: Polynomials   *26*

3-1   Polynomials   *26*
3-2   Addition of polynomials   *27*
3-3   Subtraction of polynomials   *29*
3-4   The distributive axiom   *31*
3-5   Use of parentheses and brackets   *32*
3-6   Evaluation of polynomials   *34*
3-7   Multiplication of polynomials   *35*
3-8   Division of polynomials   *37*
       Cumulative review: chapters 1 through 3   *40*

Chapter 4: Linear Equations in One Variable   *42*

4-1   What is an equation?   *42*
4-2   The solution set, *S*   *42*
4-3   Axioms which pertain to equations   *44*
4-4   Application of the addition of equals axiom   *45*
4-5   Application of the multiplication of equals axiom   *47*
4-6   More linear equations   *49*
4-7   Formulas and literal equations   *53*
4-8   Stated problems   *55*
       Cumulative review: chapters 1 through 4   *61*

*Chapter 5: Linear Inequalities   *62*

5-1   The concept of inequalities   *62*
5-2   Axioms that pertain to inequalities   *62*
5-3   The solution set of a linear inequality and its graph   *63*
5-4   Application of the axioms for inequalities   *65*

Chapter 6: Special Products and Factoring   *68*

6-1   What is factoring?   *68*
6-2   The distributive axiom   *70*

---

*Optional.

6-3   Removing a common factor   *72*
6-4   The product of the sum and difference of two terms   *73*
6-5   Factoring the difference of two squares   *75*
6-6   The product of two binomials   *77*
6-7   The square of a binomial   *79*
6-8   Factoring the quadratic trinomial   *81*
*6-9   The square of a trinomial   *84*
6-10   Summary of the vocabulary of factoring   *85*
6-11   Solving equations by factoring   *87*
        Cumulative review: chapters 1 through 6   *89*

Chapter 7: Rational Numbers (Fractions)   *91*

7-1   Various forms of the same number   *91*
7-2   Reducing fractions   *93*
7-3   Multiplication and division of fractions   *95*
7-4   Multiplication and division in the same expression   *98*
7-5   Lowest common multiple   *100*
7-6   Building fractions   *101*
7-7   Addition of fractions   *103*
7-8   Complex fractions   *108*
7-9   Evaluation of expressions containing fractions   *111*
7-10   Equations containing fractions   *113*
7-11   More stated problems   *117*
*7-12   Ratio and proportion   *120*
*7-13   Problem solving with proportion   *122*
        Cumulative review: chapters 1 through 7   *123*

Chapter 8: Graphs and Systems of Linear Equations   *125*

8-1   Linear equations in two variables   *125*
8-2   Ordered pairs   *125*
8-3   The rectangular coordinate system   *126*
8-4   The graph of a linear equation in two variables   *130*
8-5   Intercept method of graphing   *133*
8-6   Solving a system of two linear equations by graphing   *136*

---

*Optional.

8-7   Addition method of solving a system of equations   *140*

8-8   Substitution method of solving a system of
        equations   *143*

8-9   Stated problems using two equations and two
        variables   *146*

        Cumulative review: chapters 1 through 8   *148*

*Chapter 9: Linear Inequalities in Two Variables   *151*

9-1   Half-planes   *151*

9-2   The graph of an inequality in two variables   *152*

Chapter 10: Exponents and Radicals   *158*

10-1   Definitions and theorems about positive integral
        exponents   *158*

*10-2   Negative and zero integral exponents   *161*

*10-3   Scientific notation   *163*

10-4   Roots and radicals   *165*

10-5   Multiplication of radicals   *166*

10-6   Division and simplification of radicals   *168*

10-7   Addition of radicals   *171*

10-8   Further simplification of radicals   *173*

        Cumulative review: chapters 1 through 10   *175*

Chapter 11: Quadratic Equations   *177*

11-1   Definitions   *177*

11-2   An algorithm for extracting square roots   *177*

11-3   Pure quadratic equations for form $ax^2 + c = 0$ *180*

11-4   Solution by factoring—a second look   *182*

11-5   Solution by completing the square   *184*

11-6   The quadratic formula   *187*

11-7   A brief look at complex numbers   *190*

11-8   Quadratic equations with complex or imaginary
        roots   *192*

---

*Optional.

11-9   The Pythagorean theorem   *193*

11-10   Stated problems   *197*

Appendix A: Five Sets of Cumulative Review Problems and Sample
           Final Examination   *199*

Appendix B: Table of Squares, Square Roots, and Prime
           Factors   *210*

Appendix C: Answers to Selected Problems   *211*

Index   *249*

Preface

This book is designed for students with no experience in the study of algebra. It is assumed that the reader has a good understanding of ordinary arithmetic.

The main feature of *Elementary Algebra* is the cumulative review problems that appear in the daily work. Algebra is made up of many small, simple topics. Taken one at a time, these topics are easily understood. However, when several ideas are presented at one time, the subject becomes confusing. To overcome this confusion, it is necessary to work a variety of problems each day. By doing this, old material will remain fresh in the student's mind. Students using this text will not become bored by doing thirty or forty of the same type of problem. Practically every exercise is preceded by examples to illustrate the new material of that exercise. But review problems are sandwiched in between the new material in the exercises to keep the student alert. Each exercise must be completely finished to achieve the desired effect. For this reason, each exercise is the appropriate length for one day's work.

At the end of most chapters there is a special set of cumulative review problems, covering all topics studied in the text thus far. In addition, at the end of the text there are five sets of cumulative review problems that cover every nontrivial section of the entire text. A sample final examination is also included.

Much of the language of modern mathematics is used and emphasized without becoming rigorous. Every idea is presented as an axiom, theorem, or definition. The vagueness of the words "law" and "rule" is carefully avoided.

Very few proofs of theorems are given, and when they are, it is to illustrate the use of the axioms. The arbitrary development of the set of real numbers is done to emphasize that mathematics is an artistic creation, and not just a "tool of science."

Sets are introduced in chapter 1 and are used throughout the book as solution sets to all equations and inequalities. Set builder notation is used where appropriate and emphasized particularly in chapter 8.

The entire text can be covered in one semester. For a quarter system schedule, omit chapters 5 and 9 and parts of chapters 10 and 11.

The author is indebted to Professors James E. Keisler, Gerald S. Lieblich, Richard Spangler, Robert V. Kester, and William Reynolds, and to Dr. Jerald T. Ball, for their review of the original manuscript. Their suggestions led to many improvements. Special thanks also go to my wife, Dorothy, for her patient and expert typing of the manuscript, and also to Bruce Baily and Janet Greenblatt, editors for Dickenson Publishing Company, for their help in preparing this text.

P.J.M.

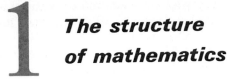

1 The structure
of mathematics

The material present in section 1-1 contains the whole heart of mathematics. It is intended that the student read this section many times during the semester and discuss its contents with the instructor. After several rereadings the student should understand the essence of mathematics.

1-1. MATHEMATICAL SYSTEMS

In this book we will be studying a *mathematical system*. The student is familiar with most of the operational aspects of this sytem, that is, the arithmetic computations he has done during previous school years.

The emphasis now turns to the structure of this system as a model for all mathematical systems. The ingredients will be introduced and explained one at a time to give the student some insight into what constitutes a creation of mathematics. The idea is to begin with nothing and carefully develop a structure in which all parts fit without contradiction; that is, the structure shall be *logically* constructed.

The system referred to above is called the *real number system*. Its first components shall be a few *undefined terms*; as the name implies, no attempt will be made to give definitions to these words. Two of these undefined terms are "addition" and "equals." (Although certain definitions will be given from time to time, they *are not a necessary part of the course structure*, but are usually convenient to have.)

Next, a small number of statements will be formulated about these undefined terms which will be accepted and adhered to without any justification. They will be accepted as true and called *axioms*. One of our axioms is the statement which allows 3 + 4 = 4 + 3.

The final necessary item will be statements concerning the undefined terms which will follow *reasonably* from the axioms. (By "reasonably" it is meant that these statements can be proved true using the axioms as justification of their truth. The actual proof will not usually be undertaken, as this is extremely difficult and subject matter for advanced courses in algebra.) This final group of statements will be called *theorems*. For example, the statement "3 × 0 = 0" is a theorem.

In short, the *axioms* are accepted without proof, whereas the *theorems* can be proved.

1-2.   UNDEFINED TERMS

Terms commonly used in this book without definition are *number*, *set*, the operation of *addition*, the operation of *multiplication*, *equal*, *point*, *line*, *between*. Descriptions and many properties of the first two terms are now given.

1-3.   THE SET OF REAL NUMBERS

In arithmetic there are two general types of numbers, whole numbers and fractions (both common and decimal). These form only a part of the set of numbers necessary for the study of algebra. Let us construct arbitrarily the set of the numbers of algebra (to be called the *real numbers*). Draw a straight line with no beginning or end, that is, indefinitely long (see Figure 1-1). Choose some point at random and give it the name zero.

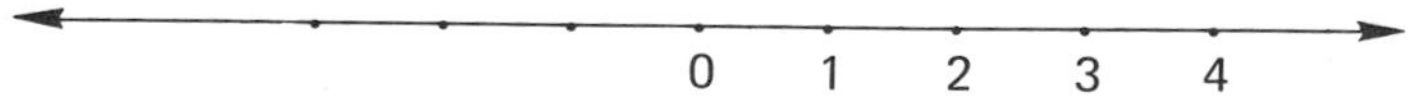

```
              0   1   2   3   4
```

FIGURE 1-1

Now choose some convenient length as a spacing device and identify the points to the right and left of the point called zero that are the same distance apart. Taking the points to the right of zero, assign the names of the whole numbers 1, 2, 3, and so forth.

Graphically it can be seen that there is space between the points named so far. Conceptually, a point has no size, so that there is no limit to the number of points between any two points. It is possible to find the point halfway between 0 and 1 and name it ½; similarly, points can be found for ⅓, ¼, ⅖, ¹⁰⁄₇, and so on, until a point has been found for every common fraction. This would seem to give names to all points between any two points. However, that is not true. Pythagoras (circa 580-500 B.C.) found that he could construct a line segment (see Figure 1-2) that was equal in length to no common fraction, and this was proved. Such a number is $\sqrt{2}$, which means the number that can be multiplied by itself to obtain a product of 2.

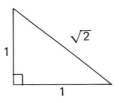

FIGURE 1-2 If each short side of the right triangle is equal to one unit, then the long side is of length $\sqrt{2}$.

If one non-fraction like $\sqrt{2}$ can be constructed, so can an infinite number of this type of number, for example, $2 \times \sqrt{2}$, $18 \times \sqrt{2}$, $\sqrt{3}$, $\sqrt{10}$, but not $\sqrt{4}$ or $\sqrt{16}$ (see Exercise 1, problem 2).

Finally, to fill in points not already named by whole numbers, fractions, or non-fractions are the nonalgebraic numbers (properly called transcendental numbers) like π, approximately 3.14159, and e, approximately 2.71. This last group will also be classified with non-fractions.

So far, all of the identified points have been to the right of the point designated as zero. Now move to the points to the left of zero and identify them in the same manner as those on the right. However, these new points will be designated as negative numbers and shall have a negative sign slightly raised and in front of the number. (See Figure 1-3.)

$$\xleftarrow{\qquad \overset{\bullet}{} \quad \overset{\bullet}{} \quad \overset{\bullet}{} \quad \overset{\bullet}{} \quad \overset{\bullet}{} \quad \overset{\bullet}{} \quad \overset{\bullet}{} \quad \overset{\bullet}{} \qquad}\longrightarrow$$
$$^-4 \quad ^-3 \quad ^-2 \quad ^-1 \quad 0$$

FIGURE 1-3

In summary, there are three distinct types of numbers which make up the set of real numbers: the positive and negative whole numbers with zero, the fractions, and the non-fractions. Each one of these sets of numbers will now be given formal mathematical names by which they will be referred to in the future.[†]

SET OF REAL NUMBERS

Common Name	Mathematical Name	Examples
positive and negative whole numbers with zero	integers*	$0, 17, 28, \ ^-4, \ ^-2$
common fractions, including decimals (positive and negative)	rational numbers	$\frac{1}{4}, \frac{84}{13}, \frac{^-1}{2}, \frac{^-5}{3}$ $.25, \ ^-.5, \ ^-1.66$
non-fractions (positive and negative)	irrational numbers	$\sqrt{2}, \ ^-\sqrt{8}, \ 6\sqrt{10}, \pi$

*The integers form a subset of the rational numbers, as they can be written in the form $^2/_1, ^3/_1$, and so on. However, the integers are usually classified separately because of their common usage.

Exercise 1

1. Explain in your own words why there is no limit to the number of points on a line between any two points.
2. Why are $\sqrt{4}$, $\sqrt{16}$, and $\sqrt{25}$ not non-fractions? (Be sure you understand the meaning of $\sqrt{}$.)
3. (a) List ten integers.
   (b) List ten irrational numbers.
   (c) List ten rational numbers that are not integers.
4. Distinguish between a theorem and an axiom.
5. Why must we begin with undefined terms?
6. What kind of real number is $1 + \sqrt{2}$?
7. Integers are placed on the number line by placing zero and adding one to get the next number. Is there a largest number? What would happen if we add one to that "largest number"?

[†] For further discussion of the real number system, see any book on the history of mathematics.

1-4. SETS

The term *set* is taken as undefined. The idea is synonymous with "a collection or group of objects." A set will be designated by a capital letter and written as follows:

$$A = \{1, 2, 3, 4\}$$

which means that A is the set whose members or elements are the numbers 1, 2, 3, and 4. It is correct to say "1 is a member of A" or "1 is contained in A"; and this will be written $1 \epsilon A$. It is also true that $2 \epsilon A$, $3 \epsilon A$, and $4 \epsilon A$. It is *not* true that $5 \epsilon A$ and we write $5 \not\epsilon A$, which is read "5 is *not* a member of A."

Consider the set $B = \{1, 2, 3\}$. It can be noted that all the members of B are also members of A, but the reverse is not true. When this occurs it is said that B "is a subset of" A, written $B \subseteq A$.

If two or more sets, say, $A = \{1, 2, 3\}$ and $B = \{3, 4, 7\}$, are to be combined, there are two common ways of doing this. One is to form a set containing all the members which are in set A or set B, or in both sets. This new set is called the union of A and B.

A union B is written $A \cup B = \{1, 2, 3, 4, 7\}$. (*Note*: Duplication of elements is avoided.)

Another way to form a set is to take only those members which are in *both* sets. We call this set the intersection of A and B. A intersect B is written

$$A \cap B = \{3\}$$

A set of particular interest in mathematics is the set which contains no members. It is called the *empty set*, or *null set*, and is symbolized by $\{\quad\}$ or $\emptyset$. The empty set is the only set which is a subset of every other set.

examples: *Given:* $A = \{1, 2, 3, 4\}$, $B = \{3, 5\}$, $C = \{2, 4\}$

   (a)   The subsets of B are $\{3, 5\}$, $\{3\}$, $\{5\}$, $\emptyset$

   (b) $A \cup B = \{1, 2, 3, 4, 5\}$

   (c) $A \cap B = \{3\}$

   (d) $C \subseteq A$

   (e) $B \cap C = \emptyset$

   (f) $A \cup C = A$

   (g) $A \cap C = C$

Exercise 2

In this exercise let $D = \{a, b, c, d\}$, $E = \{1, 2, 3\}$, $F = \{2, 3, 4, 5, 6\}$.

1. Find (a) $D \cup E$ (b) $D \cap F$ (c) $E \cup \emptyset$ (d) $F \cap \emptyset$
2. What is an axiom? (Give definition in your own words.)
3. List all subsets of E.
4. Give a description of the empty set, that is, describe a set with no members.
5. Circle the rational numbers: $\frac{3}{4}$, $^{-}\sqrt{5}$, $\frac{2}{3}$, $^{13}/_5$, $\frac{1}{6}$, $^{\pi}/_2$, 7.
6. How many empty sets are there? Why?
7. If $A \subseteq B$, what is $A \cap B$?
8. What is $A \cap \emptyset$?
9. What is $\{3, 7, 10\} \cup \emptyset$?
10. Name two undefined terms used previously in this book.
11. Why is E not a subset of F?
12. Which integers are not rational numbers?

1-5. SOME AXIOMS OF THE REAL NUMBER SYSTEM

As noted earlier, the axioms are accepted without proof. They will serve as "ground rules" for the system we are building. At this point we will accept three more undefined terms: *addition, multiplication,* and *equal.*

The following group of eleven axioms is presented *not* for memorization at this time, but to have them all in one place. In the exercises that follow throughout this book, reference is made to these axioms. It is intended that these axioms be learned through repetitive usage.

Let $R = \{$all real numbers$\}$, and let a, b, and c represent real numbers (written a, b, $c \in R$). The letters a, b, and c are called variables (see section 2-1) and are used to represent any arbitrary real numbers.

1. THE CLOSURE AXIOM FOR ADDITION: *For every pair of real numbers, a and b, there exists a unique real number, c, called the sum of a and b; that is, $a + b = c$.*

This axiom assures us that each time we add the same two real numbers, we will get the same real number for an answer.

2. THE CLOSURE AXIOM FOR MULTIPLICATION: *For every pair a, b there exists a unique real number, c, called the product of a and b, such that $a \cdot b = c$. (The dot $(\cdot)$ is used to indicate multiplication.)*

3. THE COMMUTATIVE AXIOM FOR ADDITION: $a + b = b + a$

4. THE COMMUTATIVE AXIOM FOR MULTIPLICATION: $a \cdot b = b \cdot a.$

What axioms 3 and 4 state has already been encountered in arithmetic. When learning that $2 \cdot 3 = 6$, we also learned a fact from the three times table, $3 \cdot 2 = 6$. (The word *commute* is familiar to us as meaning "to reverse the order," like commuting from home to school and back.)

5. THE ASSOCIATIVE AXIOM FOR ADDITION: $(a + b) + c = a + (b + c)$

6. THE ASSOCIATIVE AXIOM FOR MULTIPLICATION: $(a \cdot b) \cdot c = a \cdot (b \cdot c).$

Axioms 5 and 6 allow three real numbers to be added or multiplied by *grouping* them in any manner. With axioms 3 and 5 we can, for example, add a column of numbers from top to bottom or in the reverse order with the assurance that the same sum should be obtained.

Briefly, the *commutative axioms allow order changes*, while the *associative axioms allow grouping changes.*

7. THE IDENTITY FOR ADDITION: *There exists a unique real number, 0, such that for each real number a, a + 0 = a.*

8. THE IDENTITY FOR MULTIPLICATION: *There exists a unique real number, 1, such that for each real number a, a \cdot 1 = a.*

The two identity elements are sometimes called *null operators*, that is, they cause no change. When the operation $3 + 0$ takes place, the 3 remains unchanged. The term *unique*, as used in these axioms, guarantees that if 3 plus another number equals 3, then that other number must be zero.

9. THE INVERSE FOR ADDITION: *(The negative of a number) For each real number a, there exists a unique real number ^{-}a, such that a + ^{-}a = 0.*

The idea of an inverse operator is very important in mathematics. An inverse "undoes" some operation. For example, if 3 is added to a number, this operation can be "undone" by adding the negative of 3.

10. THE INVERSE FOR MULTIPLICATION: *(The Reciprocal) For each real number a, where a ≠ 0, there exists a unique real number ¼a, such that a · ¼a = 1.*

Again this axiom shows us how to obtain the identity element for a particular operation.

Axioms 9 and 10 enable us to define subtraction and division, which we shall do at a later time.

11. THE DISTRIBUTIVE AXIOM: $a \cdot (b + c) = (a \cdot b) + (a \cdot c)$.

This is one of the most frequently used axioms, and it is the only one that involves addition and multiplication. It is properly called "the distributive axiom for multiplication over addition." If you study the axiom carefully, you can see how the "a" is distributed or "handed out" to each of the addends inside the parentheses.

Exercise 3

State the axiom being used in problems 1 to 14.

examples:

(a)	$a + b = b + a$	commutative for addition
(b)	$xy + 0 = xy$	identity for addition
(c)	$x + (y + 2) = x + (2 + y)$	commutative for addition
(d)	$(ab)(cd) = ((ab)c)d$	associative for multiplication
(e)	$6(30 + 4) = (6 \cdot 30) + (6 \cdot 4)$	distributive
(f)	$16 \cdot \frac{1}{16} = 1$	inverse for multiplication

1. $a \cdot x = x \cdot a$
2. $a \cdot (x + y) = a \cdot x + a \cdot y$
3. $4 + 0 = 4$
4. $6 \cdot \frac{1}{6} = 1$
5. $8 + 5$ is always 13
6. $7 + (^-7) = 0$
7. $(a \cdot b) \cdot c = (b \cdot a) \cdot c$
8. $(a + b) + c = (b + a) + c$
9. $x + (y + z) = (x + y) + z$
10. $a \cdot (k + 2) = (a \cdot k) + (a \cdot 2)$
11. $3 \cdot (a \cdot (x + y)) = (3 \cdot a) \cdot (x + 7)$
12. $1 \cdot (a + b) = (a + b)$
13. $1 \cdot (a + b) = 1 \cdot a + 1 \cdot b$
14. $(a + x) \cdot (b + y) = (a + x) \cdot b + (a + x) \cdot y$
15. What is the inverse for addition for 7?
16. What is the inverse for multiplication for 3?
17. Give additive identity for

(a) 6   (b) $\frac{1}{3}$ (c) $\frac{^-5}{6}$

18. What is multiplicative identity for

    (a) ⁻5   (b) $\dfrac{3}{4}$ (c) 0

19. What is the reciprocal of $\dfrac{1}{3}$?

20. Give two rational numbers that are not integers.

21. Name two irrational numbers other than $\sqrt{2}$ or π.

22. What is the negative of $\dfrac{1}{4}$?

2 Four operations on the real numbers

2-1. REPRESENTING REAL NUMBERS WITH VARIABLES

It would be impossible for us to list all the real numbers each time we want to make a statement about them. At other times we may want to say something about one or a few real numbers but are not sure exactly which ones. In both of these instances a letter will be used to represent some unknown number or numbers. This letter will be called a *variable*.

DEFINITION: *A variable is a symbol used to represent an unspecified member of a set.*

All algebraic expressions will contain one or more variables together with addition, subtraction, multiplication, and division symbols. (See sections 2-5 and 2-7 for definitions of subtraction and division.)

examples: $(x + 1)$ means "one more than some number, where x represents some unknown number"

$x(x + 1)$ means "the product of some number and one more than that number"

$\dfrac{x}{y}$ means "the quotient of two numbers"

$5 + xy$ means "five more than the product of two numbers"

$n, n + 2$ These two expressions represent two consecutive even integers if n is even *or* two consecutive odd integers if n is odd.

2-2. SUMS AND TERMS

In arithmetic, when $3 + 4 = 7$ is written, the 3 and 4 are called addends; in algebra, *any numbers which are added are called terms.* Each algebraic expression contains one or more terms and the entire expression is called a *sum.*

examples: $3x^2 + 2xy + 4z$ contains three terms
$5x^2 y^2 z^2$ contains one term
$x + y$ contains two terms

2-3. FACTORS AND EXPONENTS

When writing $3 \times 4 = 12$ (or $3 \cdot 4 = 12$ or $3(4) = 12$), the 3 and 4 are called *factors*, as they are numbers which are multiplied. The single word "factor" replaces the traditional "multiplier" and "multiplicand," since expressions containing more than two factors will be written frequently.

examples: $5xyz$ contains four factors
$8(x + y)$ contains two factors, and one of these factors, $(x + y)$, contains two terms
$(x + y)(x - y)$ contains two factors

When writing an expression in which all factors are the same, such as $x \cdot x \cdot x$, we will use the notation x^3. The 3 is called the *exponent* and the x is called the *base*. The symbol x^3 is read "x cubed" or "x to the third power." Similarly, x^2 means $x \cdot x$ and is read "x squared" or "x to the second power."

examples: $(x + y)^3 = (x + y)(x + y)(x + y)$
$(xy)^2 = (xy)(xy)$
$x^6 = x \cdot x \cdot x \cdot x \cdot x \cdot x$ and is read "x to the sixth power"

DEFINITION: *If n is a positive integer then* $x^n = x \cdot x \cdot x \cdot \ldots$ *(to n factors).*

As we progress, it will become very important to be able to count the number of terms and factors that are part of an *entire* algebraic expression.

examples:

$x + y$ contains 2 terms and only one factor; that is, the entire expression itself is considered as the only factor.

$a(x + y)$ contains two factors, a and $(x + y)$, but only one term. Remember, we are considering the expression as a whole and the addition of x and y inside the parentheses does not involve the a, thus only one term.

$ax + y$ contains two terms and one factor.

$(x + y)(a + b)(c + d + e)$ contains 3 factors and only one term.

It should be noted that every expression will contain at least one term and one factor, and always only one of one type.

Exercise 4

1. In algebra, the name which replaces addend is _____.
2. In algebra, numbers which are multiplied are called _____.

Write an algebraic expression for problems 3 through 9.

3. Eight less than some number
4. Twice some number increased by 13
5. Twice the square of some number
6. Three more than one-half some number
7. The square of the sum of x and y
8. The product of the sum of 8 and k and twice some other number
9. The quotient obtained when a number that is added to twice itself is divided by 21
10. *List examples of the following:*

    (a) five rational numbers that are not integers
    (b) five negative integers
    (c) three irrational numbers

11. If $A = \{12, 15, 18, 21\}$ and $B = \{3, 7, 12, 18, 24\}$,
     what is $A \cap B$?

Give the number of terms and factors in each expression in problems 12 through 20 (refer to the examples at the end of section 2-3):

12. $xy + z$ 13. $k^2 z^2 + xyz$
14. $(x + y)(x + 2y)$ 15. $k(m + n)$
16. $x(x + y)(x^2 + y^2)$ 17. $a + bc + def$
18. $xyzw$ 19. $3m^2 + 8m + 6$
20. $a^3 + 1$
21. Represent two numbers such that one of them is 3 more than twice the other.
22. If n represents an integer, represent the next two consecutive integers.
23. If n is an *even* integer, what will the next three consecutive even integers look like?
24. If n is an odd integer, give the next four consecutive odd integers and the first odd integer which is smaller than n.
25. What is an axiom?
26. If m is twice some other number, represent that other number using m in your answer.
27. *Explain:* If n is an integer, then $2n$ is always even.
28. *Explain:* If n is an integer, then $2n + 1$ is always odd.

2-4. THE ABSOLUTE VALUE OF A REAL NUMBER

The real numbers have been pictured on a number line with zero as a central reference point, zero being the only real number which has no algebraic sign (plus or minus). Upon studying the number line, we can see that the numbers 4 and $^-4$ are the same distance from zero. This idea of "distance from zero" is called the *absolute value* of a real number. The symbol for the absolute value of x is $|x|$. The precise definition now follows:

DEFINITION: $|x| = x$ *if x is positive or zero*
$|x| = ^-x$ *if x is negative*

Distance is always positive (by choice) and the absolute value of any real number is also positive. According to the definition: $|4| = 4$ and $|^-4| = ^-(^-4) = 4$ (see Theorem 2-6).

2-5.　ADDITION OF REAL NUMBERS

All real numbers are either positive or negative, except zero, which has no sign. You must learn how to compute with these new numbers. Addition is an undefined term and we must accept it as such. However, there are several axioms that tell us in which order to add.

We can think of addition as movement along the real number line:

1. Movement to the right will indicate addition of positive numbers.
2. Movement to the left will indicate addition of negative numbers.

Assuming that all addition problems begin at zero, consider the following problems:

$^+4 + {}^+3$　means move 4 units to the right of zero, then 3 more units to the right. The sum is $^+7$.

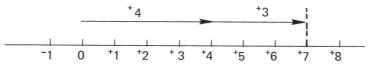

$^-4 + {}^-3$ means move 4 units to the left of zero, then 3 more units to the left. The sum is $^-7$.

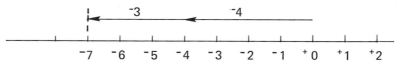

$^-4 + {}^+3$　means move 4 units to the left of zero, then 3 units to the right of $^-4$. The sum is $^-1$.

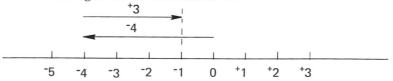

$^+4 + {}^-3$ means move 4 units to the right of zero, then 3 units to the left of $^+4$. The sum is $^+1$.

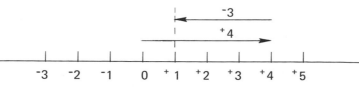

examples: *Add the following.*

(a) ⁻8	(b) ⁻2	(c) ⁺3	(d) ⁺65	(e) ⁺134
⁻3	0	⁻5	⁻17	⁺216
⁻11	⁻2	⁻2	⁺48	⁺350

After studying these examples, a generalization is in order. In examples (a) and (e) above, each sum was obtained by adding the absolute values of each number and prefixing the common sign. In the other three examples, the sum is the arithmetic difference between the absolute values of each term, and the sign of the number with the larger absolute value is the sign of the sum.

When more than two terms are to be added, the following theorem may be used.

THEOREM 2-1: *(The Rearrangement Theorem for Addition)*
A number of terms may be rearranged in any order before addition is performed.

Justification of this theorem is made by repeated use of the commutative and associative axioms for addition.

For example: ⁻3 + ⁺4 + ⁺2 + ⁻5 + ⁻6 may be rearranged as
⁻3 + ⁻5 + ⁻6 + ⁺4 + ⁺2.

Now sum the negative terms: ⁻3 + ⁻5 + ⁻6 = ⁻14
Then sum the positive terms: ⁺4 + ⁺2 = ⁺6

Finally, add ⁻14 + ⁺6 = ⁻8, which is the final sum.

examples: (a) $\lceil 5 \rceil = {}^-(\,{}^-5) = 5$

(b) $|2| = 2$

(c) $\lceil 6 + {}^-2 \rceil = |{}^-8| = {}^-(\,{}^-8) = 8$

(d) $\,{}^-|4| = {}^-4$

(e) ⁺2 + ⁻4 + ⁺3 + ⁺6 + ⁻5

    $= ({}^+2 + {}^+3 + {}^+6) + ({}^-4 + {}^-5)$

    $= {}^+11 + {}^-9$

    $= {}^+2$

(f) Add: ⁻2⌐
      ⁺3
    ⁺8 ⌊⁺5 ⌉ ⁻8
      ⁻6⌐

    ⁺8 + ⁻8 = 0

Exercise 5

Perform these algebraic additions:

1. $^+8$	2. $^-4$	3. $^-17$	4. $^+11$	5. $^-18$
$^+3$	$^-3$	$^-13$	$^-5$	$^+27$

6. $^-8$	7. $^-2$	8. $^+6$	9. $^-53$	10. $^+64$
$^-3$	$^+5$	$^-4$	$^-21$	0
$^-11$	$^+7$	$^+4$	$^+34$	$^-18$
		$^-6$		

11. State the commutative axiom for addition.
12. State the associative axiom for addition and give a numerical example to illustrate this axiom.

Find the following sums:

13. $^-6 + ^-3$
14. $^+8 + ^-5$
15. $^-3 + ^-8 + ^-4 + ^+6$
16. $^+3 + ^-4 + ^-2 + ^-5$
17. $^-18 + ^+13 + ^+15 + ^-4$
18. $^-6 + ^-8 + ^-3 + ^-6$
19. How many terms are there in problem 15? How many factors?
20. Name two irrational numbers.

Add:

21. $^-3 + ^-4 + ^+6 + ^-2 + ^-5 + ^-6 + ^+2 + ^+5$
22. $^+6 + ^+7 + ^-3 + ^+4 + ^-8 + ^-3 + ^-5 + ^+5$
23. $^-7 + ^+7 + ^-3 + ^-3 + ^+5 + ^-2 + ^-7 + ^-3$
24. $^+41 + ^-31 + 0 + ^-5 + ^-16 + ^+3 + ^-4 + ^-15$
25. $|6| + |\ 7| + \lceil 3|$
26. $|3 + ^-4| + \lceil 8 + ^-9|$
27. $^-|4| + ^-\lceil 5| + ^-3 + 7$
28. $\lceil 3 + ^-5 + 2 + 3| + \lceil 6 + ^+6$

2-6. SUBTRACTION OF REAL NUMBERS

The idea of subtraction or "take away" is familiar to us from arithmetic. However, this operation is not mentioned in the axioms for algebra. Let us look at some examples to see how subtraction should be introduced into algebra.

We know 6 – 4 = 2.

In cold weather the temperature might be 4° and suddenly drop 6°. The new temperature would be 2° below zero, which is written $^-2°$.

This situation suggests the subtraction $4 - 6 = {}^-2$, a problem not possible in arithmetic but handled easily with signed numbers.

Finally, suppose you are \$3 in debt (you have ${}^-3$ dollars). If someone "takes away" that debt you are suddenly richer by \$3. This suggests that taking away (subtracting) a negative number is the same as adding the same positive number.

Now let us see if we can discover a uniform connection between these three subtraction problems and the operation of addition.

$$6 - 4 = \ 2 \text{ could be written } 6 + {}^-4 = \ 2$$
$$4 - 6 = {}^-2 \text{ could be written } 4 + {}^-6 = {}^-2$$

Subtracting negative 3 is the same as adding ${}^+3$,
that is, $- {}^-3 = + {}^+3$.

These observations lead us to define subtraction as follows:

DEFINITION OF SUBTRACTION: $a - b = a + {}^-b$
In words: To subtract b from a means to add the negative of b to a.

examples: (a) ${}^+8 - {}^+3 = {}^+8 + {}^-3 = {}^+5$

              (b) ${}^+8 - {}^-3 = {}^+8 + {}^-({}^-3) = {}^+8 + {}^+3 = {}^+11$

              (c) ${}^-8 - {}^-3 = {}^-8 + {}^-({}^-3) = {}^-8 + {}^+3 = {}^-5$

Examples (b) and (c) require the following theorem.

THEOREM 2-2 ("THE DOUBLE NEGATIVE" THEOREM):   *If $a \epsilon R$, then ${}^-({}^-a) = {}^+a$.*
In words: The additive inverse of the additive inverse of a real number is the number itself.

Another important theorem.

THEOREM 2-3 ("THE NEGATIVE OF A SUM" THEOREM):   *If a, $b \epsilon R$, then ${}^-(a + b) = {}^-a + {}^-b$.*

In words: The negative of a sum is the sum of the negatives of each term.

The following examples use both theorems and the definition of subtraction:

examples: (a) ${}^-(x + {}^-y) = {}^-x + {}^+y$

              (b) ${}^-(a + b - c) = {}^-(a + b + {}^-c) = {}^-a + {}^-b + {}^+c$

              (c) ${}^+6 + {}^-4 - 2 - {}^-3 + {}^-5 \ = {}^+6 + {}^-4 + {}^-2 + {}^+3 + {}^-5$

                            $= {}^+9 + {}^-11 = {}^-2$

(d) $(^+a - ^+b) - (^+c + ^+d) = (a + ^-b) + ^-(^+c + ^+d)$
$$= a + ^-b + ^-c + ^-d$$

(e) $(^+a + ^-b) - ^-(^+c + ^-d) = (^+a + ^-b) + (c + ^-d)$
$$= ^+a + ^-b + ^+c + ^-d$$

Exercise 6

Subtract the bottom number from the top:

examples: (a) $^+4$ means $^+4$ (b) $^-2$ means $^-2$
              $^+3$ $+\ ^-3$ $^-5$ $+\ ^+5$
              ——   ———   ——   ———
                                 $^+1$ $^+3$

1. $^+8$ 2. $^-8$ 3. $^-16$ 4. $^-12$ 5. $^+15$
    $^+6$ $^+3$ $^+4$ $^-2$ $^-15$
    ——   ——   ———   ———   ———

6. $^-24$ 7. $^+50$ 8. $^+43$ 9. $^+6$ 10. $^+52$
    $^+13$ $^-20$ $^+51$ $^-100$ $^+70$
    ———   ———   ———   ————   ———

11. Explain subtraction in your own words.
12. State the distributive axiom and illustrate with a numerical example.
13. What is a theorem?
14. Do problems 1 to 10 in this exercise, but add the terms instead of subtracting.

Follow the operation signs:

15. $12 - ^-2$ 16. $40 + ^-8$ 17. $^-24 - ^+15$
18. $(20 + ^-4) - ^-3$ 19. $^-(^+4 + ^-2) + ^-6$ 20. $^+7 - (8 - ^-2)$
21. What is the additive inverse of 6? of $^-4$? of $^-\frac{1}{3}$?
22. If two numbers, a and b, are added and the sum has the same algebraic sign as the sign of b, then what is true about a and b?

Follow the operation signs:

23. $^+3$ 24. $^-8$ 25. $^+13$ 26. $^-21$ 27. $^-18$
     $+\ ^-4$ $-\ ^-2$ $-\ ^+8$ $+\ ^-5$ $-\ ^-3$
     ———   ———   ———   ———   ———

28. $^-40$ 29. $^+6$ 30. $^-4$ 31. $^-28$ 32. $^+16$
      $^-10$ $-\ ^-4$ $+\ ^-6$ $-\ ^-28$ $+\ ^-3$
    $+\ \ 5$
    ———   ———   ———   ———   ———

Note: For the remainder of this book, no algebraic sign will appear on a positive number, so that $^+6$ will be written 6.

2-7. MULTIPLICATION OF REAL NUMBERS

Multiplication, like addition, is an undefined term. The following argument motivates the theorems concerning multiplication of signed numbers.

With the aid of the number line, note the movement of successive products:

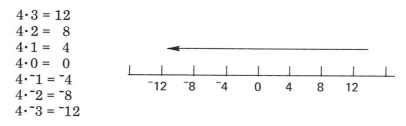

4·3 = 12
4·2 = 8
4·1 = 4
4·0 = 0
4·⁻1 = ⁻4
4·⁻2 = ⁻8
4·⁻3 = ⁻12

Note that 4·3 and 4·⁻3 are additive inverses of each other.

Now follow the movement of these products:

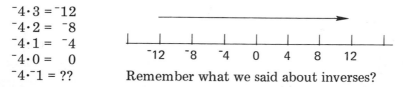

⁻4·3 = ⁻12
⁻4·2 = ⁻8
⁻4·1 = ⁻4
⁻4·0 = 0
⁻4·⁻1 = ??   Remember what we said about inverses?

Note that the products are moving to the right (in a positive direction). We will conclude that the product of ⁻4 and ⁻1 is 4.

Continuing:

⁻4·⁻1 = 4
⁻4·⁻2 = 8
⁻4·⁻3 = 12

In the following theorems, *a and b represent positive real numbers.*

THEOREM 2-4: $a \cdot b = ab$

THEOREM 2-5: $^{-}a \cdot b = {}^{-}ab$

THEOREM 2-6: $^{-}a \cdot {}^{-}b = ab$

There are many other intuitive ways of explaining the theorems for multiplication of signed numbers; but the only acceptable argument is the proof of each of the above theorems.

examples:
(a) $3 \cdot 5 = 15$
(b) $^-3 \cdot 5 = ^-15$
(c) $3 \cdot ^-5 = ^-15$
(d) $^-3 \cdot ^-5 = 15$
(e) $(^-2)^2 = (^-2)(^-2) = 4$
(f) $^-(2)^2 = ^-(2)(2) = ^-4$
(g) $^-2^2 = ^-(2)(2) = ^-4$

Observe very carefully that $(^-2)^2 \neq ^-2^2$.

Exercise 7

Find the product in each problem:

1. 6	2. $^-7$	3. 4	4. 12	5. $^-2$
3	4	$^-8$	5	$^-3$

6. $^-11$	7. 12	8. 20	9. $^-5$	10. $^-13$
$^-5$	6	$^-3$	$^-8$	4

11. 50	12. 6	13. $^-5$	14. 8
10	$^-7$	$^-4$	$^-4$

15. $(^-3)^2$　　16. 1^2　　17. 8^2　　18. $(^-4)^3$
19. $^-(3^2)$　　20. $^-(^-3)^2$　　21. $^-(^-5)(2)$　　22. $(^-7)(8)$
23. $^-6(5)$　　24. $(^-11)(^-8)$　　25. $8(^-7)$　　26. $(^-14)(2)$
27. $-\left(\frac{1}{3}\right)\left(^-\left(\frac{1}{2}\right)\right)$　　28. $3\left(^-\left(\frac{1}{8}\right)\right)$　　29. $^-\left(\frac{3}{16}\right)\left(\frac{2}{9}\right)$　　30. $^-\left(\frac{5}{6}\right)\left(-\left(\frac{3}{4}\right)\right)$

31. Why could you predict that problems 22 and 25 would have the same answer?
32. For problems 1 to 14, find the algebraic sums.

Find each product:

33. $(.02)(^-.3)$　　　34. $(.3)^2$　　　35. $(^-.01)^2$
36. $(^-.001)(^-.02)$　　　37. $(3½)^2$　　　38. $(^-2½)^3$
39. $(4.3)(^-.01)(^-.3)$　　　40. $(^-1)^{40}$

2-8.　DIVISION OF REAL NUMBERS

Division, like subtraction, is defined into existence. The axiom for multiplicative inverses assures us that each real number a, other than zero, has a unique real number associated with it called ¼a. Using this axiom, we state the definition of division.

DEFINITION OF DIVISION: $a \div b$ or $\dfrac{a}{b}$ means $a\left(\dfrac{1}{b}\right)$, $b \neq 0$

In words: To divide a by b means to multiply a by the multiplicative inverse (reciprocal) of b, b not equal to zero.

examples:

(a) $6 \div 3 = 6\left(\dfrac{1}{3}\right) = 2$

(b) $\dfrac{8}{-4} = 8\left(\dfrac{1}{-4}\right) = {}^-2$

(c) $8\overline{)\,{}^-16} = {}^-16\left(\dfrac{1}{8}\right) = {}^-2$

It should be observed that the algebraic signs for quotients are the same as for products, as division is a form of multiplication.

examples:

(a) $\dfrac{{}^-3}{-3} = 1$

(b) $\dfrac{6}{2} = 3$

(c) $\dfrac{{}^-6}{2} = {}^-3$

(d) $\dfrac{6}{-3} = {}^-2$

Exercise 8

Divide:

1. $\dfrac{16}{4}$

2. $\dfrac{21}{{}^-3}$

3. $\dfrac{11}{{}^-1}$

4. $\dfrac{{}^-6}{{}^-3}$

5. $\dfrac{{}^-18}{3}$

6. $\dfrac{{}^-28}{4}$

7. $\dfrac{{}^-35}{{}^-7}$

8. $\dfrac{40}{{}^-8}$

9. $\dfrac{52}{13}$

10. $\dfrac{{}^-60}{20}$

11. $\dfrac{365}{{}^-5}$

12. $\dfrac{{}^-4250}{{}^-10}$

Follow the operation signs:

13. ${}^-21 + {}^-3$
14. $({}^-21)({}^-3)$
15. $16 - ({}^-2)$
16. Subtract ${}^-11$ from ${}^-21$
17. $8 + {}^-4 - {}^-2 + 3$
18. $({}^-3)^2\,({}^-2)({}^-5)$
19. $24 \div ({}^-5)$
20. ${}^-50 \div ({}^-5)$
21. Explain division in words.

2-9. ORDER OF PERFORMING OPERATIONS

Many albegra and arithmetic problems involve more than one operation. Even a simple problem like $2 + 3 \cdot 4$ is ambiguous if no order is observed. One person might say $2 + 3 \cdot 4 = 5 \cdot 4 = 20$, while some-one else may see it as $2 + 3 \cdot 4 = 2 + 12 = 14$. Obviously we cannot have such confusion. The following order is observed universally in mathematics and science.

Order of Performing Operations:

First:　　Perform operations which are inside parentheses.
　　　　Note: The numerator and denominator of a fraction are considered to be in a parentheses.

Second:　Raise all numbers to powers as indicated by exponents.

Third:　　Perform multiplications and divisions in order of appear-ance from left to right.

Fourth:　Perform additions and subtractions in order from left to right.

examples:　　(a)　$3 + 4 \cdot 2 = 3 + 8 = 11$

(b)　$\dfrac{3 + 4}{3} = \dfrac{7}{3} = 2\dfrac{1}{3}$

(c)　$5^2 + 2 = 25 + 2 = 27$

(d)　$3 \cdot (4 + 2) = 3 \cdot 6 = 18$

(e)　$5^3 \cdot 6 + 5 = 125 \cdot 6 + 5$
　　　　　　　　　$= 750 + 5 = 755$

(f)　$\dfrac{18 + 4^2}{3 - 20} = \dfrac{18 + 16}{^{-}17}$
　　　　　　　$= \dfrac{34}{-17} = {}^{-}2$

Exercise 9

1.　$2 \cdot 3 + 4 \cdot 6$
2.　$8({}^{-}6) - 2({}^{-}4)$
3.　$4^2 - 5^2$
4.　$(3(6) + 2 \cdot 7)^2$
5.　$\dfrac{8 - 2}{^{-}3 - 9}$
6.　$2^3 \cdot 3 - 1 \cdot 4 + 6$
7.　$8(3 + 2) - 2(6 + 1)$
8.　$16 \div ({}^{-}4) + 2 \cdot 5^2$
9.　$3(2 + (3 \cdot 4 - 1) - 6)$
10.　$5 \cdot (2 + 4) - 3(2 + 1)$
11.　State the commutative axiom for addition and give a numerical example.
12.　By definition, $16 \div 5$ means _____ .
13.　What is the additive inverse of 6?
14.　What is the multiplicative identity?

15. $6 + 2 \cdot 4 - 8^2 = ?$
16. $11^2 - 2^2 + 5 \cdot 4 \cdot 3 \div 2 = ?$
17. Name three different kinds of real numbers.
18. What is the sum of 40 and ⁻50?
19. What is the product of 12 and the square of 20?
20. What is the square of the sum of 8 and 12?

2-10.   EVALUATION OF ALGEBRAIC EXPRESSIONS

Algebraic expressions containing variables are used to represent real numbers. Sometimes, particular real numbers are known for the variables. When this occurs, it is possible to give a particular value to an expression. This process is called *evaluation of an expression*.

examples: Let $a = 1,$ $b = ^{-}2,$ $c = 3$

(a) $a + b$ becomes $1 + (^{-}2) = ^{-}1$
(b) $2a - 3b$ becomes $2(1) - 3(^{-}2) = 2 + 6 = 8$
(c) $2b^2$ becomes $2(^{-}2)^2 = 2 \cdot 4 = 8$
(d) $(2b)^2$ becomes $(2(^{-}2))^2 = (^{-}4)^2 = 16$
(e) $a^2 - bc - c^2$ becomes $(1)^2 - (^{-}2)(3) - (3)^2$
$$= 1 - (^{-}6) - 9$$
$$= 1 + 6 - 9 = ^{-}2$$

Exercise 10

Evaluate the following: If $x = 2,$ $y = ^{-}3,$ $z = ^{-}1$

1. $3x$ 2. xy 3. $xy + z$
4. xyz 5. $2y - z$ 6. z^2
7. y^2 8. $3y^2$ 9. $x(x + y)$
10. $x^2 + y^2$ 11. $8z + yz$ 12. $x^2 y^2 z^2$
13.   Take a few minutes and reread section 1-1.
14.   Give a definition of a theorem.
15.   What is the algebraic sign of the product of eleven negative factors and three positive factors?
16.   Which real number has no multiplicative inverse?

Evaluate the following: If $a = 3,$ $b = ^{-}1,$ $c = ^{-}2$

17. $4a + 2b + 3c$ 18. $\dfrac{a + b}{c}$

19. $\dfrac{a}{a^2 + b^2}$

20. $3 + \dfrac{ac}{b}$

21. $ab + ac - bc$

22. $2ab + 3ab$

23. $\dfrac{c}{a + 3b}$

24. $ac - 3c^2 + 5b^2$

25. State the associative axiom for multiplication and illustrate with a numerical example.
26. State the distributive axiom.
27. We know $16 \cdot 1 = 16$ because of the _____ axiom.
28. Give an example of a real number whose square is negative.
29. Distinguish between the "square of the sum of x and y" and the "sum of the squares of x and y."
30. Which is larger, $^-8$ or $^-12$?

Cumulative Review: Chapters 1 and 2

1. Name three undefined terms which have been used in this book.
2. The three necessary parts of a mathematical system are _____ , _____ , and _____ .
3. The real numbers 2 and $\sqrt{5}$ belong to two distinct subsets of R. What are the names of these two sets?
4. If $A = \{8, 12, 16, 20\}$ and $B = \{12, 20, 28\}$, find
    (a) $A \cap B$ (c) $A \cap \emptyset$
    (b) $A \cup B$ (d) $B \cup \emptyset$
5. What is
    (a) the reciprocal of $\frac{1}{3}$?
    (b) the multiplicative inverse of $\frac{1}{3}$?
    (c) the additive identity for $^-6$?
6. State the associative axiom for addition.
7. Subtract $^-19x^2 y^2$ from $21x^2 y^2$.
8. Give an expression which contains exactly three terms and one factor.
9. What is $1 - 3 + {}^-2 - 51$?
10. *Simplify:*
    (a) $(xy)(3xy^2)$ (c) $24x^3 \div 8x$
    (b) $(2xy)(xy) - 3(xy^2)x$ (d) $2x^2 - 3x^3 + 4x^2$
11. Use the distributive axiom to multiply $3(40 + 7)$.
12. Define division of real numbers.
13. *Simplify:*
    (a) $^-4(3) + 2(6)(^-3) - 15$
    (b) $16 \div 2 + 2 - 4 \div 4$

14. If $c = 3$ and $d = {}^-2$, evaluate
   - (a) $c^2 d^2$
   - (b) ${}^-2d^2$
   - (c) $({}^-2d)^2$
   - (d) $4cd - 8cd + 6cd^2$
   - (e) $(c - d)^2$
   - (f) $(c + d)(c - d)$

15. In your own words, describe the importance of having both axioms and theorems in a mathematical system.

3 *Polynomials*

3-1. POLYNOMIALS

In this course all algebraic expressions will be called *polynomials.** This word is a blend of Greek and Latin and translates literally as "many names (of numbers)." The "many" refers to the number of terms in each expression; and each polynomial receives a special name according to its number of terms. To illustrate:

$$3x \quad \text{has one term and is called a } monomial$$
$$x + 2y \quad \text{has two terms and is called a } binomial$$
$$x + y + z \quad \text{has three terms and is called a } trinomial$$

Each of these above expressions is also called a polynomial.

Polynomials are classified according to the *number of variables* and the *degree*, which is the largest sum of the exponents occurring in any term. Thus:

$$x + y \quad \text{is a first-degree polynomial in two variables}$$
$$x^2 + x + 3 \quad \text{is a second-degree polynomial in one variable}$$
$$x^2 + y^2 \quad \text{is a second-degree polynomial in two variables}$$
$$xy + 3x \quad \text{is a second-degree polynomial in two variables}$$

A more familiar name for "first degree' is *linear*; and for "second degree" is *quadratic*.

---

*In a more advanced course in algebra a distinction is made between algebraic expressions that are polynomials and others called rational expressions. This distinction is not necessary for this book.

In any monomial there are two types of factors, called *coefficients*. In the monomial $3x$, 3 is the *numerical coefficient* and x the *literal coefficient*. If two terms have identical literal coefficients, they are said to be *similar terms*.

examples: (a) $3x$ and $2x$ are similar terms
  (b) $3x$ and $3y$ are *not* similar terms
  (c) $3x^2 y$, $^-16x^2 y$ and $4.7x^2 y$ are similar terms
  (d) $5x^2 y$ and $5xy^2$ are *not* similar terms

Exercise 11

Give degree, number of variables, and name of each polynomial:

1. $2x$ 2. $5x + 3$ 3. $x + y$
4. $5x^2 + 3x$ 5. $2x^3 + 3x + 2$ 6. $8x^2 + 7x + 8$
7. $x^2 + y^2 + z^2$ 8. $4x^6 - 7$ 9. $x - 2$
10. $3x^2 + 3$ 11. $2x^4 + x^2 + 4$ 12. $x^3 + y^2 + x$
13. Define "similar terms."
14. For each term give three similar terms:
     (a) $11y$ (b) $2k^2$ (c) $21m^3 n^2$
     (d) $^-43x^2 y^2$ (e) 21
15. What is the difference between an axiom and a theorem?
16. $21 + {}^-3 - 2 + {}^-4 = ?$ 17. $18({}^-2) + 3({}^-5) = ?$
18. $24 \div (3 + {}^-2) = ?$ 19. Subtract $^-18$ from 24.
20. $5 - ({}^-8) + 6({}^-4)({}^-3) = ?$ 21. $21 \cdot 6 \div 2 + 3 = ?$
22. Illustrate the commutative axiom for addition.
23. Illustrate the distributive axiom.
24. Is $\sqrt{2}$ a real number?

3-2. ADDITION OF POLYNOMIALS

It may seem reasonable to say that $3x + 2x = 5x$; or that $7xy - 5xy = 2xy$. Both of these statements are correct and can be justified by using the distributive axiom and some basic addition facts.

Proof that $3x + 2x = 5x$:

$3x + 2x = x \cdot 3 + x \cdot 2$ by commutative axiom for multiplication
    $= x(3 + 2)$ by distributive axiom
    $= x(5)$ by addition of real numbers
    $= 5x$ by commutative axiom for multiplication

THEOREM 3-1: $ax + bx = (a + b)x$, *where a and b are* usually *known numbers (constants).*

The above theorem and examples allow the following conclusion:

Terms may be combined (added) into a single term if and only if they are similar.

It is impossible to justify adding terms which are not similar.

examples: (a) $12x + {}^-2x = 10x$
              (b) $8y + 2z + 3y = 11y + 2z$
              (c) $2xy^2 + 3x^2y + 4xy^2 + {}^-8xy^2 = {}^-2xy^2 + 3x^2y$
              (d) $3x + 4y + 5z$
                      $8x + {}^-2y + {}^-3z$
                  $+ \ 7x + {}^-5y + 6z$
                  $18x - 3y + 8z$

From these examples it can be seen that similar terms are added by *adding their numerical coefficients* while keeping the literal coefficients the same.

Exercise 12

Add:

1. $3x$	2. $5xy$	3. ${}^-7k^3z$	4. $3(x + y)$
$2x$	${}^-2xy$	${}^-2k^3z$	${}^-4(x + y)$
${}^-4x$	${}^-6xy$	$2k^3z$	${}^-2(x + y)$

5. $6x + 3x - 2y - 5y$
6. $10x^2 + 2x + 3x^2 - 5x$
7. $8x + 3 + 2x - 4 - 6 + 2x$
8. $21m^2 - {}^-16m^2 + 3m^2 + 5m$
9. $52x^3y + 3xy - 6xy + 11x^3y$
10. $15x^2 + 21x + 3$

11. $3a - 2b - \ c - \ d$ 12. $3xy + 2cd$
    ${}^-2 + \ b - \ c + d$ $5xy - 3cd$
    $6a - 3b + 2c - 3d$ ${}^-4xy - 5cd$

13. $3a^2 + 2b^2 - \ z^2$ 14. $3m + 4n - 2c - 3d$
      $a^2 \qquad - \ z^2$ $6m - 7n - 3c + 7d$
    $4a^2 - 2b^2 + 3z^2$

15. $3axy + 4$
     $5axy - 6$
     $^-6axy + 3$

16. $c - d$
     $2a + 4c$
     $3a \quad\quad - 2d$

Perform the indicated operations:

17. $(^-3)^2$ 18. $(^-3)(4)$ 19. $(^-3)(^-5)$ 20. $(^-3)^3(4)$
21. What is the additive identity?
22. What is the additive inverse of $^1/_6$?
23. *Add:* $4x$, ^-3k, ^-3x, and $5k$.
24. *Add:* $3x^3$, $2x^2$, $5x^3$, $^-7x^3$, $^-18x^2$, $3x$, and 4.

3-3. SUBTRACTION OF POLYNOMIALS

Recall that the subtraction $a - b$ is defined as $a + (^-b)$. The question now is "What is the additive inverse of $(a + b)$?" The answer is given in the following theorem.

THEOREM 3-2: $^-(a + b) = ^-a + ^-b$
In words: The negative of a sum is the sum of the additive inverse (negative) of each term.

To subtract $(2a + b)$ from $(5a + 2b)$ means to add the negative of $(2a + b)$ to $(5a + 2b)$.

That is,

$(5a + 2b) - (2a + b)$
$= (5a + 2b) + ^-(2a + b)$ by definition of subtraction
$= 5a + 2b + ^-2a + ^-b$ by Theorem 3-2
$= 3a + b$ by addition of similar terms
         Theorem 3-1

examples:

(a) $(2ax - 4ay) - (3ax - 5ay) = (2ax + ^-4ay) + ^-(3ax + ^-5ay)$
$= 2ax + ^-4ay + ^-3ax + 5ay$
$= ^-ay + ay$

In a more abbreviated form this problem becomes

$(2ax - 4ay) - (3ax - 5ay) = 2ax - 4ay - 3ax + 5ay$
$= ^-ax + ay$

(b) $(3x^3 - 4x^2 + 6x + 2) - (5x^3 + 2x - 3)$
$= 3x^3 - 4x^2 + 6x + 2 - 5x^3 - 2x + 3$
$= ^-2x^3 - 4x^2 + 4x + 5$

Sometimes subtraction problems are given in a vertical arrangement, such as:

$$3x - 4y + 2$$
$$-\ \ (6x + 5y - 3)$$

In this case we appeal to the definition of subtraction again and "add the negative of the subtrahend." Very simply, the negative of a polynomial is a polynomial with the same terms but opposite signs. So the above problem becomes

$$3x - 4y + 2$$
$$+\ \ ^-6x - 5y + 3$$

It may be inconvenient to rewrite a subtraction problem to show the subtrahend with new signs. In this case, a common practice is to show the sign changes by writing the new signs *above* the old signs.

$$\begin{array}{c} 3x - 4y + 2 \\ -\ \ 6x + 5y - 3 \end{array} \quad \text{becomes} \quad \begin{array}{c} 3x - 4y + 2 \\ +\ \mp\ 6x \mp 5y \pm 3 \\ \hline ^-3x - 9y + 5 \end{array}$$

This last procedure is strongly recommended.

Exercise 13

Give the negative (additive inverse) of each polynomial:

1. $3x + 2$ 2. $x + y$ 3. $5x - y$
4. $^-x + 2$ 5. $3x^2 y + 4xy - 3$ 6. $x^3 + y^3 + z^3$
7. xy 8. $^-3x^2 - 2x + 4$ 9. $21m^3 + 6m^2 - 8m - 2$

10. Define subtraction: $a - b$ means _____.

Subtract the lower polynomial:

11. $3x^2 + 3x + 6$ 12. $3x - 3y$ 13. $a + b + c$
    $5x^2 - 3x - 6$ $^-4x - 3y$ $a + b - c$

14. $^-8xy + 3x + 4$ 15. $5m^2\ \ \ \ \ + 3$ 16. $\ \ \ abc +\ ef$
    $^-2xy - 3x + 6$ $2m^2 - 3m + 5$ $^-8abc - 2ef$

Using Theorem 3-2, perform the following operations:

17. $(5x + 2) - (6x - 3)$ 18. $(3x + 5) - (2x + 1)$
19. $(5k^2 + 3k + 2) - (6k^2 + 5k - 4)$ 20. $(5 - x) - (6 - x)$
21. $(3xy - y - 4) - (2xy - 3)$
22. $(5x^2 + 3x + 1) - (2x^2 + 5x - 3)$

23. Name three different types of real numbers.
24. What special mathematical name does 1 have.
25. If four negative numbers are added, is the sum positive or negative?

3-4. THE DISTRIBUTIVE AXIOM

The distributive axiom is used quite frequently and deserves special attention.

It states:

$$a(b + c) = ab + ac$$

examples: (a) $3(x + 4) = 3x + 12$

(b) $2(x - 6) = 2x - 12$

(c) $^-5(x + y + z) = ^-5x - 5y - 5z$

Example (c) illustrates the *generalized distributive theorem*, which allows us to distribute the factor outside the parentheses over *each term* on the inside, no matter how many terms there are.

Exercise 14

Perform each multiplication by using the distributive axiom or theorem and simplify:

1. $4(x + y)$
2. $3(x - 5)$
3. $^-6(x - y - z)$
4. $2(x^2 + y^2 - z^2)$
5. $12(x^3 - x^2 + x - 3)$
6. $14(a - b - c + 3)$
7. $1(3a + 2b + 5)$
8. $^-1(x - y - z)$
9. $8(k^2 z + kz - 4)$
10. $3(x - 2) + 6(x + 4)$
11. $5(m^2 + m - 4) - 3(m^2 + 6)$
12. $6(a^2 + a + 3) - 11(a^2 + a + 2)$
13. $2(x + y) + 3(x - y) + 5(x - 2)$
14. $^-7(x^2 y^2 + z) - 8(x^2 y^2 + 3)$
15. $5(abc + 2ab + a) + 3(abc + a)$

Give the number of terms and factors in each expression:

16. $3x^2 y - 2xy$
17. $4(x + 2y)$
18. $(5 + x)(x - 3)$
19. $2(x + 3) - 4y$
20. $x^2 y^2 + xy^2 + xy^3$

3-5.   USE OF PARENTHESES AND BRACKETS

Parentheses, ( ), and brackets, [ ], are used for purposes of grouping. Both symbols are used in the same way.

To remove parentheses, consider the following theorems:

THEOREM 3-3: $(a + b) = a + b$

Proof: $(a + b) = 1 \cdot (a + b)$ multiplicative identity axiom

$\qquad\qquad = 1 \cdot a + 1 \cdot b$ distributive axiom

$\qquad\qquad = a + b \qquad$ multiplicative identity axiom

Also:

THEOREM 3-4: $^-(a + b) = {}^-a + {}^-b$

This theorem was given and used in section 3-3 and can be proved
in exactly the same manner as Theorem 3-3.

If each theorem in this section is read from right to left, then we
can see how to group terms inside parentheses.

examples: (a) $5x + 2y + 3 = (5x + 2y + 3)$

(b) $3x + y - 4 = {}^-({}^-3x + {}^-y + 4)$
$\qquad\qquad\qquad = {}^-({}^-3x - y + 4)$

(c) $^-8x - 4 = {}^-(8x + 4)$

Examples (b) and (c) can be stated:

To group terms within parentheses or square brackets
preceded by a negative sign: change the sign of each term
and then place inside the grouping symbol.

If an expression contains parentheses within parentheses, or parentheses within brackets, then the recommended method for simplification is to remove the innermost parentheses first.

examples: (a) $[a + b + (2a - b)] = [a + b + 2a - b] = [3a] = 3a$

(b) $^-[x - (2x - y) + (3x + y)] = {}^-[x - 2x + y + 3x + y]$
$\qquad\qquad\qquad\qquad\qquad = {}^-[2x + 2y] = {}^-2x - 2y$

Exercise 15

Simplify:

1. $(2x + y + (3x - y))$
2. $(a - (a + b) - (a - b))$
3. $[a + 2(a + b) - 3(a + b)]$
4. $2[x + 3(x - y)]$
5. $4[2(a + b) - 3(a - b)]$
6. $2a - (a + 4) + 16 + (3a - 6)$
7. $p + 3q + 2[p - (2q + 3)] + 3p$
8. $a - b - [a - b - (a - b) - (a - b)]$
9. $x + 3 + 2(x + 4) - 3(x - 1) + {}^-(x - 4)$
10. ${}^-(x + 2) - (3x - 8) + (4x + 2) - 3(3x - 1)$
11. How many factors are there in this expression:
    $(3x + y)(ab + c)$?
12. $3({}^-4 + 2) + 2({}^-6) = ?$
13. (a) $16 \div ({}^-4) = ?$ (b) $({}^-6)({}^-4) = ?$ (c) ${}^-6 + {}^-4 = ?$
14. Represent the sum of the squares of two numbers.
15. Give four undefined terms.

In the following problems, group the first three terms in a parentheses preceded by a positive sign and the remaining terms in a parentheses preceded by a negative sign:

16. $2a - 3x + 5y + 2z + 3$

17. ${}^-xy - 2z + 3 - 7x + 5y - 3$

18. $21k + 6k^2 + 4k^3 - 4k^4 + 3k^5 + 2k^6$

19. $x - y - z + w - u + p$

20. $a^2 + 4a + 4 - x^2 + y^2$

Use parentheses to write the following expressions:

21. The sum of a and b is *subtracted from* 6.
22. The sum of x and y is *added to* the sum of k and 4.
23. The difference obtained, when $2y$ is subtracted from $3x$, is subtracted from m.
24. Three more than the sum of x and y is added to the difference obtained when k is subtracted from 8.
25. Define subtraction in words.
26. If $A = \{9, 10, 11, 12\}$ and $B = \{10, 12, 14, 16\}$, what is
    (a) $A \cup B$
    (b) $A \cap B$
27. What are two symbols used to represent the empty set?
28. $\{a, b, c\} \cap \{1, 2, 3\} = ?$

3-6. EVALUATION OF POLYNOMIALS

One way of checking the simplification of a polynomial is to evaluate it both before and after simplification.

example: If $x = 2$, $y = 1$, $z = {}^-2$,   evaluate

$$3x^2 + 2y - 4x^2 + z - 3z = {}^-x^2 + 2y - 2z$$

$$3(2)^2 + 2({}^-1) - 4(2)^2 + ({}^-2) - 3({}^-2) \quad ? \quad {}^-(2)^2 + 2({}^-1) - 2({}^-2)$$
$$3\cdot4 - 2 - 4\cdot4 - 2 + 6 \ ? \ {}^-4 - 2 + 4$$
$$12 - 2 - 16 - 2 + 6 \ ? \ {}^-2$$
$${}^-2 = {}^-2$$

Exercise 16

Evaluate each polynomial before and after simplification:

Let $a = 3$, $b = 1$, $c = {}^-2$

1. $2a + b - a + 5$
2. $(a + c) - (a - b - c)$
3. $2(a + b) - 3(a + b)$
4. $a^2 + b^2 + (2a^2 + 3b^2 + c^2)$
5. $a - b - 2a + c + 3c - (4a + c)$
6. $5(c - a) + 2(a + b) - 3(c + b)$
7. $2a^2 b^2 + 3a^2 b^2 - 4a^2 b^2 + 7a^2 b^2$
8. $2[3(a + c) - 4(2(a + b)) - 5(c + a)]$
9. ${}^-[{}^-(a - b) + (2a + 3b) - 4(a + c)]$
10. $3ac - 5ab + 4bc - 2ac + 6ab - ac$
11. Illustrate the commutative axiom for addition.
12. Which axiom is this: $a + (b + c) = (a + b) + c$?
13. Which axiom assures us that once we learn the times tables, they will never change?
14. Use the distributive axiom to complete this problem:
    $4\cdot52 = 4(50 + 2) = $?
15. Use the distributive axiom to complete the following:
    (a) $5\cdot65$ (b) $7\cdot28$ (c) $8\cdot346$

3-7.   MULTIPLICATION OF POLYNOMIALS

If more than two factors are to be multiplied, then the associative axiom for multiplication allows us to group the factors in any manner we choose. By the commutative axiom for multiplication, any order of factors may be attained. Thus, we may state the following theorem:

THEOREM 3-5: *(The Rearrangement Theorem for Multiplication)*
If a real number contains several factors, then these factors may be rearranged in any order.

To apply this theorem, consider the following operations:

$$(3xy)(6xy) = 3 \cdot x \cdot y \cdot 6 \cdot x \cdot y = (3 \cdot 6)(x \cdot x)(y \cdot y) = 18x^2 y^2$$

Also:

$$(2xyz)(3x)(yz)(2y) = (2 \cdot 3 \cdot 2)(x \cdot x)(y \cdot y \cdot y)(z \cdot z)$$
$$= 12x^2 y^3 z^2$$

Or:

$$(^-xy)(3x^2)(2y^2)(5) = (^-1 \cdot 3 \cdot 2 \cdot 5)(x \cdot x^2)(y \cdot y^2)$$
$$= {}^-30x^3 y^3$$

Note: Multiplication has taken place *without* requiring similar terms for factors.

To multiply polynomials having more than one term, the distributive axiom is used one or more times.

examples:

(a) $3x(2x + 3y) = 3x \cdot 2x + 3x \cdot 3y = 6x^2 + 9xy$

(b) $(x + y)(a + b) \ = \ (x + y)a + (x + y)b$ distributive axiom

$\qquad\qquad\qquad = \ a(x + y) + b(x + y)$ commutative axiom
$\qquad\qquad\qquad\qquad\qquad\qquad\qquad$ for multiplication

$\qquad\qquad\qquad = \ ax + ay + bx + by$ distributive axiom

In example (b), four terms appear in the product. If these terms are examined carefully, you can see that each term in the first binomial is multiplied by each term in the second binomial. This type of problem lends itself to vertical multiplication similar to that learned in arithmetic.

examples: Multiply.

(a) $\qquad$ $3x + y$
$\qquad\qquad$ $2x - y$
$\qquad\qquad$ $\overline{- 3xy - y^2}$
$\qquad$ $\underline{6x^2 + 2xy}$
$\qquad$ $6x^2 - xy - y^2$

(b) $\quad$ $2x + 3y - 2z$
$\qquad\quad$ $x + 2$
$\qquad\quad$ $\overline{4x + 6y - 4z}$
$\qquad\qquad\qquad - 2xz + 3xy + 2x^2$
$\qquad$ $\overline{4x + 6y - 4z - 2xz + 3xy + 2x^2}$

Exercise 17

Multiply and simplify if possible:

1. $2ax(3x)$
2. $4xy(^-5xy)$
3. $m(mn)$
4. $m^2 n^2 (2mn^2)$
5. $abc(3ab)$
6. $(^-6x)(3x)(4xy)$
7. $5k(2k)(k)$
8. $xy(3x)(2x^2)(x^2 y^2)$
9. $^-4(^-2x)(^-2y)(^-3x)$
10. $5ab(2ab)(3b^2)(a)$
11. $2x^2 + 3x(6) + 5x^2$
12. $2ax(x) + 3a(5x^2) - 4x(ax)$
13. $5a(b^2) + 2a^2 (b) + 3ab(b) - 6a(ab)$
14. $2x^2 (x) + 3x(4x^2) + 5x^2 (^-2x) + 7x(x)(x)$
15. $3np(2n^2) - 4n^2 (np) - 5np(n) + n^2 p^2$
16. What is the difference between an axiom and a theorem?
17. State the commutative axiom for multiplication.
18. $xy + xy + xy + xy = ?$
19. From $3x^2 y$ subtract $11x^2 y$.
20. If $A = \{2, 3, 4, 5\}$, is $5 \notin A$?

Multiply:

21. $3x + 2$
$\quad$ $\underline{2x - 5}$

22. $5x + 3$
$\quad$ $\underline{x - 2}$

23. $2m^2 + 1$
$\quad$ $\underline{5m + 1}$

24. $8x + y$
$\quad$ $\underline{2x - y}$

25. $m^2 + 3m + 1$
$\qquad\quad$ $\underline{m + 1}$

26. $3x^2 + 2x + 4$
$\qquad\quad$ $\underline{x - 5}$

27. $x^2 - 2x + 4$
$\qquad\quad$ $\underline{x + 2}$

28. $2x^2 + 3x + 1$
$\quad$ $\underline{4x^2 - 2x - 3}$

29. $^-3a + 2a - 5$
$\quad$ $\underline{a^2 + 3a + 6}$

30. $(x + 2)(x - 3) + 2x^2 + 7$
31. $(x - 4)(x - 2) + 3(x - 4)$
32. $4x^2 - 3(x^2 + 7) - 2(x + 6)$
33. What special mathematical name is given to the real number zero?
34. $^-5(x + y - 3) + 2(3x + 4y - 5) + 2(y - 6) = ?$

35. If $x = {}^-3$, what is the value of ${}^-4x^3$?
36. *Define division:* $a \div b$ means _____
37. *Find each quotient:*

    (a) $\dfrac{{}^-36}{4}$ (b) $\dfrac{{}^-18}{-2}$ (c) $\dfrac{10}{-5}$

38. $(x + 3)(x + 3) - (x + 3)(x - 3) = ?$

3-8. DIVISION OF POLYNOMIALS

Division problems fall into two categories, depending upon the number of terms in the divisor. If the divisor is a monomial, a process similar to short division is employed.

examples: (a) $\dfrac{6x}{2} = 3x,$ Check: $3x(2) = 6x$

                 (b) $\dfrac{5x^2}{x} = 5x,$ Check: $5x(x) = 5x^2$

                 (c) $\dfrac{8x^3}{2x} = 4x^2,$ Check: $4x^2(2x) = 8x^3$

To justify these solutions:

    (a) $\dfrac{6x}{2} = \dfrac{2 \cdot 3x}{2 \cdot 1} = \dfrac{2}{2} \cdot \dfrac{3x}{1} = 1 \cdot \dfrac{3x}{1} = 3x$

    (b) $\dfrac{5x^2}{x} = \dfrac{5x \cdot x}{1 \cdot x} = \dfrac{5x}{1} \cdot \dfrac{x}{x} = \dfrac{5x}{1} \cdot 1 = 5x$

    (c) $\dfrac{8x^3}{2x} = \dfrac{2 \cdot x \cdot 4x^2}{2 \cdot x \cdot 1} = \dfrac{2}{2} \cdot \dfrac{x}{x} \cdot \dfrac{4x^2}{1} = 1 \cdot 1 \cdot 4x^2 = 4x^2$

Division of a polynomial by a monomial will be treated in section 7-2.

If the divisor is a polynomial, then the algorithm for division is used. Recall from arithmetic.

$$
\begin{array}{r}
4 \\
12\,\overline{\big)\,563} \\
48 \\
\hline
83
\end{array}
$$

1. Divide 56 by 12

2. Multiply 4 by 12

3. Subtract 48 from 56
4. Bring down next number

This completes one cycle in the division process.

Now consider this example:

$$\begin{array}{r} x \hspace{2.5em} \\ x+2\,\overline{\smash{\big)}\,x^2 - 3x - 10} \\ \underline{x^2 + 2x} \hspace{2.5em} \end{array}$$

1. Divide x^2 by x
2. Multiply x by $(x + 2)$

$$\begin{array}{r} x \hspace{2.5em} \\ x+2\,\overline{\smash{\big)}\,x^2 - 3x - 10} \\ \underline{\mp\, x^2 \mp 2x} \hspace{2.5em} \\ -5x - 10 \end{array}$$

3. Subtract $x^2 + 2x$ from $x^2 - 3x$
4. Bring down ($^-10$)

This completes one cycle. Now begin again.

$$\begin{array}{r} x \;\; -5 \hspace{1em} \\ x+2\,\overline{\smash{\big)}\,x^2 - 3x - 10} \\ \underline{\mp x^2 \mp 2x} \hspace{2.5em} \\ -5x - 10 \\ \underline{\pm\, 5x \pm 10} \\ 0 \end{array}$$

1a. Divide (^-5x) by x
2a. Multiply ($^-5$) by $(x + 2)$
3a. Subtract
4a. Nothing left to "bring down"

The above example illustrates a problem which has no remainder.

Let us try $(x^2 + 4x + 5) \div (x + 2)$

$$\begin{array}{r} x \;\; +2 \hspace{1em} \\ x+2\,\overline{\smash{\big)}\,x^2 + 4x + 5} \\ \underline{\mp x^2 \mp 2x} \hspace{2.5em} \\ 2x + 5 \\ \underline{\mp 2x \mp 4} \\ 1 \end{array}$$

The remainder is 1, and it should be incorporated into the quotient by placing it over the divisor and adding this fraction to the partial quotient already obtained. Thus, the entire quotient becomes $x + 2 + \dfrac{1}{x + 2}$.

There is one point of procedure when using this algorithm: the dividend and the divisor must have their terms ordered so that exponents are in descending order. The division problem $x + 3\,\overline{\smash{\big)}\,6x - 2x^2 + 7}$ must be rewritten $x + 3\,\overline{\smash{\big)}\,{}^-2x^2 + 6x + 7}$. Moreover, if we have the problem $x + 1\,\overline{\smash{\big)}\,x^3 + 1}$, the rule concerning descending order of exponents still must hold. Since the polynomial $x^3 + 1$ has no x^2 or x term, a space for them must be left in the dividend.

It is best to write $x^3 + 1$ as $x^3 + 0x^2 + 0x + 1$.

$$
\begin{array}{r}
x^2 - x + 1 \\
x + 1 \overline{\smash{\big)}\ x^3 + 0x^2 + 0x + 1} \\
\underline{\pm x^3 \pm x^2} \\
- x^2 + 0x \\
\underline{\mp x^2 \mp x} \\
x + 1 \\
\underline{\mp x \mp 1} \\
0
\end{array}
$$

Exercise 18

Perform the indicated operations:

1. $12x^2 (3x)$
2. $8x(^-2x)$
3. $\dfrac{8x}{-2x}$
4. $\dfrac{12x^2}{3x}$
5. $\dfrac{^-18xy}{2x}$
6. $5x(3x - 2)$
7. $^-6y(^-3y + 4)$
8. $\dfrac{120x^2}{6x}$
9. $\dfrac{14mk}{2k}$
10. $\dfrac{^-24x^2 y}{^-6xy}$

11. Subtract 42 from $^-18$.
12. $3(^-2)^2 - 2^2 = ?$
13. What is the reciprocal of $^-3$?
14. (a) $16x^2 \div 4x^2$ (b) $28k \div 14k$ (c) $38k(2k)$
15. (a) $16x^2 + 4x^2$ (b) $28k - 14k$ (c) $38k \div 2k$
16. (a) $3x(x + 2)(x + 3)$ (b) $4(x + 2)(x - 2)$
17. Give the definition of a set.
18. Enclose in parentheses preceded by a negative sign $3x - 2y - 2z + 3$.

Use the long division algorithm:

19. $x + 1 \overline{\smash{\big)}\ x^2 + 4x + 3}$
20. $x - 2 \overline{\smash{\big)}\ x^2 + 6x - 8}$
21. $x + 4 \overline{\smash{\big)}\ x^2 + 3x - 4}$
22. $k - 6 \overline{\smash{\big)}\ 4k^2 - 21k + 2}$
23. $(8a^2 - 6a - 5) \div (2a + 1)$
24. $(6m^2 - 17m - 3) \div (m - 3)$
25. $(x^2 + 3xy - 4y^2) \div (x + 2y)$
26. $(x^2 - 2x + 3) \div (x + 5)$

27. $(a^3 - 1) \div (a - 1)$
28. $(m^3 - 9m^2 + 26m - 24) \div (m - 2)$
29. $(2a^4 + 7a^3 + 6a^2 - 7a - 7) \div (a^2 + a - 2)$
30. $(6m^3 - 5m + 8m^4 + 8m^2 - 3) \div (4m^2 - 1 - m)$

Simplify:

31. $20x^3(3x) \div 6x^2$

32. $\dfrac{5x^2 + 2x^2}{7}$

33. $\dfrac{4x(x + 3x)}{^-2}$

34. $\dfrac{4m^2 + 3m^2 - 2m^2}{3m}$

35. $\dfrac{3(^-3)^2 + 4(2)^2 + 1}{3(^-5) + (^-7)}$

Cumulative Review: Chapters 1 through 3

1. Give an example of a second-degree polynomial in three variables.
2. What kind of positive real number is ¾?

Simplify:

3. $4(x - 3) + 2(2x + 1) - 3(x - 8)$
4. (a) $3(x + 2)(x - 3)$ (b) $(y + 1)^2$
5. $2[3x - (2x + y) - y(3) + 5x]$
6. $^-[8((x - y) + 2(y - x)) + 4]$
7. $(x + y + z) - (3x + 2y + z) + (5x - 8y - 10z)$
8. (a) $(3x^2 - 2x + 1)(2x - 3)$ (b) $x - 3 \overline{\smash{\big)}\, x^2 + 9x - 36}$
9. From $9x^2 - 3y^2$ subtract $5x^2 + 8y^2$.
10. (a) Give a numerical example to illustrate the distributive axiom.
    (b) Use the associative axiom for addition to add 3, 4, and 8 in two different ways.
11. *Simplify:*
    (a) $2y(3x - 4y) - 6x(3y - x) - (^-x^2)$
    (b) $(3x)^2 - 3x^2 - (^-3x)^2 - 3^2 x^2$
12. Use the associative axiom for multiplication to do this problem in two different ways: $3(y + 2)(y + 3)$.
13. Use the distributive axiom for each problem.
    (a) $5y(xy + yz)$ (b) $(5y + 2)(3y - 1)$
14. If $a = 3$, $b = 0$, and $c = ^-2$, evaluate:
    (a) $4a^2 - b^2$ (b) $b^3 - c^3$
    (c) $(ab)^3 + (ac)^2$ (d) $^-3ac - c^2 + (3c)^2$

15. Give another name for "linear" and for "second degree."
16. Distinguish between an axiom and a theorem. What do they have in common?
17. Show a number whose negative is a positive number.
18. Give the following unique real numbers:
    (a)   the reciprocal of 21
    (b)   the additive inverse of ⁻18
    (c)   the absolute value of ⁻24
    (d)   the additive identity
    (e)   the multiplicative inverse of 21
    (f)   the multiplicative identity
19. *Give an example of*
    (a)   a negative irrational number
    (b)   a positive rational number, larger than 4, that is not an integer
20. Given $A = \{5, 10, 15, 20, 25\}$ and $B = \{0, 10, 20, 30\}$, find:
    (a) $A \cap B$ (b) $A \cup B$ (c) $B \cap \emptyset$

4

Linear equations in one variable

4-1. WHAT IS AN EQUATION?

The word *equals* is undefined. An *equation* is a statement that two numbers (polynomials) are equal. An equation may be true, such as $3 + 4 = 7$; it may be false, such as $3 + 2 = 9$; or it may be neither true nor false, such as $x + 3 = 8$. This latter type is called a *conditional equation*, because it is neither true nor false.

Equations are classified in the same manner as polynomials, that is, according to degree and number of variables. Recalling that *linear* and *first-degree* are synonomous, the equation $x + 3 = 8$ *is called a linear equation in one variable.* In chapter 8 we will study linear equations in two variables, such as $x + y = 7$; and in chapter 11, quadratic (second-degree) equations in one variable, such as $x^2 + 3x + 2 = 0$. But for the present we will concentrate on linear equations in one variable.

4-2. THE SOLUTION SET, *S*

The main objective of this chapter is to take conditional equations and make them true. In the equation $x + 3 = 8$, x gets the usual designation as a variable, that is, it can be replaced with any real number. If we let $x = 5$ in the above equation, then the conditional equation $x + 3 = 8$ becomes the true equation $5 + 3 = 8$. This value for x which makes the equation true is called the *root* or *solution* of the equation. For this type of equation, any other value for x would make the equation false and is of no interest to us.

The root or roots of an equation will be placed in a set which we will call the *solution set* of the equation, and this special set is denoted by S. Every equation that we solve in this text will have its roots placed in a solution set.

Now that sets will be used quite frequently, some new notation should be introduced. The set $A = \{1, 2\}$ could be written

$$A = \{x \mid x = 1 \text{ or } x = 2\}$$

This form is called *set builder notation* and is read "A is the set of real numbers x such that x equals 1 or x equals 2." The vertical bar after the first x is read "such that."

Similarly,

$$B = \{x \mid x + 2 = 5\}$$

In this case, B is the solution set to the equation $x + 2 = 5$. It is read "B is the set of real numbers x such that $x + 2 = 5$ is true." You might also note that in this case B has only one number, 3; and $B = \{3\}$.

Exercise 19

1. What is an equation?
2. Give an equation which is not true.
3. Give an equation which is true.
4. *Give an example of*
    - (a) a linear equation in one variable
    - (b) a linear equation in three variables
    - (c) a third-degree equation in one variable
    - (d) a quadratic equation in one variable
5. If $x + 5 = 9$, would $S = \{6\}$ be the solution set? Why?
6. Is $S = \{2\}$ the solution set for the conditional equation $x + 8 = 10$? Why?
7. State the associative axiom for multiplication.
8. What is the multiplicative inverse of ¾?
9. *List the members of*
    - (a) $A = \{x \mid x = 4 \text{ or } x = 6 \text{ or } x = 2\}$
    - (b) $B = \{x \mid x = 2 \text{ or } x = 5\}$
    - (c) $C = \{x \mid x = 3 \text{ and } x = 5\}$
    - (d) $D = \{x \mid x \text{ has no members}\}$
    - (e) $E = \{x \mid x \text{ is an odd integer between 4 and 18}\}$

(f) $F = \{x \mid x$ is the 37th President of the United States$\}$
(g) $G = \{x \mid x$ is the *number* of humans on the Sun$\}$
(h) $H = \{x \mid x$ is the human on the Sun$\}$

10. $x + 3 \overline{\smash{\big)}\, x^2 + 8x + 15} = ?$
11. $^-4(^-3)(^-1)(2) + ^-18 \div 3 = ?$
12. Find the value of $3x^3 + 2x^2 - x - 3$ if $x = ^-1$.
13. Name three general types of real numbers.
14. What is an axiom?
15. What is a solution set?
16. If $A = \{x \mid x + 5 = 9\}$, is
    (a) $4 \epsilon A$?
    (b) $6 \epsilon A$?
17. Use the distributive axiom to multiply: $3x^2 (4x^2 + 2x)$
18. How many terms and factors are there in the following expressions? (Refer to section 2-3.)
    (a) $xy + xz + xw$ (d) $(y + 2)(y + 3)$
    (b) $x(x + y)$ (e) $(y + 2)(y + 3) + y + 3$
    (c) $a + a(b + c)$
19. Find $\{x \mid x = 3$ or $x = 4$ or $x = 5\} \cap \{x \mid x$ is an even integer$\}$.
20. Find $\{x \mid x$ is an odd even integer$\} \cup \{x \mid x = 6\}$.

4-3.   AXIOMS WHICH PERTAIN TO EQUATIONS

In order to solve an equation (find its solution set, S), some axioms must be taken. In the following axioms, a, b, and c are real numbers:

Reflexive axiom: $a = a$

Symmetric axiom: If $a = b$, then $b = a$

Transitive axiom: If $a = b$ and $b = c$, then $a = c$

Addition of
equals axiom: If $a = b$, then $a + c = b + c$

Multiplication of
equals axiom: If $a = b$, then $ac = bc$

The first three axioms are given merely to complete the list of axioms pertaining to equations. No attempt should be made to memorize them. The last two axioms are fundamental to the solution of equations and will now be used extensively.

For example: The addition of equals axiom states that if

$$3 = (2 + 1) \text{ then}$$
$$3 + 4 = (2 + 1) + 4 \text{ by adding 4 to each side of the equation}$$
$$7 = 3 + 4$$
$$7 = 7$$

The multiplication of equals axiom states that if

$$4 = 6 - 2 \text{ then}$$
$$^-3(4) = {}^-3(6 - 2) \quad \text{by multiplying each side by } {}^-3$$
$$^-12 = {}^-18 + 6$$
$$^-12 = {}^-12$$

4-4. APPLICATION OF THE ADDITION OF EQUALS AXIOM

According to the addition of equals axiom, any positive or negative real number may be added to both sides of an equation without destroying the equality. The choice of the number or numbers to be added to both sides will depend upon the numbers which are already in the equation.

Consider:
$$x + 5 = 8$$
$$(x + 5) + {}^-5 = 8 + {}^-5 \quad \text{Add } {}^-5 \text{ to both sides}$$
$$x + 0 = 3$$
$$x = 3$$
$$S = \{3\}$$

The plan for solving any equation is to set up a series of operations, using the axioms, which will result in an equation with the variable occurring on one side by itself. There are two important axioms that tell us when a real number can be written "by itself."

The additive identity axiom: $a + 0 = a$
The multiplicative identity axiom: $a \cdot 1 = a.$

examples: (a)   Find the solution set, S:

$$3x - 2 - 2x = 1$$
$$x - 2 = 1 \qquad \text{Combine similar terms}$$
$$x - 2 + 2 = 1 + 2 \quad \text{Add 2 to both sides}$$
$$x = 3$$
$$S = \{3\}$$

Check: If $x = 3$
$$3x - 2 - 2x = 1$$

becomes

$$3(3) - 2 - 2(3) = 1$$
$$9 - 2 - 6 = 1$$
$$9 - 8 = 1$$
$$1 = 1 \quad \text{which is true.}$$

(b)   Find S:

$$\begin{aligned}
5x + 2 - (3x - 2) &= x + 6 \\
5x + 2 - 3x + 2 &= x + 6 \\
2x + 4 &= x + 6 \\
2x + 4 + {}^-4 &= x + 6 + {}^-4 \quad \text{Add } {}^-4 \text{ to each side} \\
2x &= x + 2 \\
2x + {}^-x &= x + 2 + {}^-x \quad \text{Add } {}^-x \text{ to each side} \\
x &= 2 \\
S &= \{2\}
\end{aligned}$$

(c)   Find S:

$$\begin{aligned}
2(x + 1) - 3x + 5 &= 4x - 3 - (6x + 5) \\
2x + 2 - 3x + 5 &= 4x - 3 - 6x - 5 \\
{}^-x + 7 &= {}^-2x - 8 \\
{}^-x + 7 + 2x &= {}^-2x - 8 + 2x \quad \text{Add } 2x \text{ to each side} \\
x + 7 &= {}^-8 \\
x + 7 + {}^-7 &= {}^-8 + {}^-7 \quad\quad\quad \text{Add } {}^-7 \text{ to each side} \\
x &= {}^-15 \\
S &= \{{}^-15\}
\end{aligned}$$

Exercise 20

Find the Solution Set, S, by using the addition of equals axiom:

1. $x - 3 = 2$
2. $x + 6 = 5$
3. $x - 3 = {}^-3$
4. $x - 8 = 4$
5. $x + 7 = 9$
6. $x + \frac{1}{3} = \frac{2}{3}$
7. $x - 5 = {}^-3$
8. $x - 4 = 4$
9. $x + 6 = {}^-6$
10. $x - 7 = 0$
11. State the addition of equals axiom.
12. Find the sum of 8, ${}^-6$, 3, 4, ${}^-2$, 5, ${}^-1$.
13. Find the product of ${}^-2$, 3, ${}^-1$, ${}^-1$, 2.
14. Find S, if $x + 3 = 21$.
15. What is the degree and number of variables in the polynomial $5x^3 + 2x^2 - x + y$?

Find S, combining similar terms first:

16. $x + 3 - 4 = 16 - 3$
17. $3x + 4 - 2x = 7 - 18$
18. $6x + 3 = 5x - 4$
19. $21m + 5 = 18m + 3 + 2m$
20. $2k - 3 - 8 - (k + 1) = 0$
21. $3(x + 4) + 2(x + 3) - (4x + 6) = 5$
22. $2z + 4z - 8 = 5 - z - 2z$

23. $^-(x + 4) - 3x + 5(x - 3) = 1$
24. $2(x - 6) - 3(3 - x) = 2(x + 2) + 2x - 3$
25. $y + 2(y - 8) = 5(y + 4) + 3(3 - y)$
26. If $x = 3$ and $y = {}^-2$, what is the value of $3xy^2 + 2xy$?

4-5. APPLICATION OF THE MULTIPLICATION OF EQUALS AXIOM

The multiplication of equals axiom allows us to multiply both sides of an equation by any real number without destroying the equality. The application of this axiom becomes a matter of choosing the appropriate number for each equation. If an equation were in the form $1 \cdot x = 3$, then the solution becomes obvious. All we must do then is to transform the coefficient of x into the number one. The multiplicative inverse (reciprocal) axiom shows us how to do this.

examples: (a) Find S:

$$3x = 6$$ Choose ⅓ as the factor for each side of the equation (⅓ is chosen because it is the reciprocal of 3).

$$⅓(3x) = ⅓ \cdot 6$$
$$1 \cdot x = 2$$
$$x = 2$$
$$S = \{2\}$$

(b) Find S:

$$^-8x = 12$$
$$^-(⅛)(^-8x) = {}^-(⅛)(12)$$
$$x = {}^-(^{12}\!/_8) = {}^-(^3\!/_2)$$
$$S = \{^-(^3\!/_2)\}$$

(c) Find S:

$$^3\!/_5\, x = ½$$
$$^5\!/_3(^3\!/_5\, x) = {}^5\!/_3(½)$$
$$x = ^5\!/_6$$
$$S = \{^5\!/_6\}$$

Recall that division is defined in terms of multiplication. Accordingly, the multiplication of equals axiom permits division on both sides of an equation by the same real number.

Examples (a), (b), and (c) above can also be done by dividing both sides of the equation by the same number, as follows:

examples: (a) $3x = 6$

$$\frac{3x}{3} = \frac{6}{3}$$

$$x = 2$$

$$S = \{2\}$$

(b) $^-8x = 12$

$$\frac{^-8x}{^-8} = \frac{12}{^-8}$$

$$x = {}^-\left(\frac{3}{2}\right)$$

$$S = \left\{{}^-\left(\frac{3}{2}\right)\right\}$$

(c)   This problem can be done in two steps.

$$\frac{3}{5}x = \frac{1}{2}$$

$$5\,\frac{3}{5}x = 5\left(\frac{1}{2}\right) \qquad \text{Multiply both sides by 5}$$

$$3x = \frac{5}{2}$$

$$\frac{3x}{3} = \frac{5}{2} \div 3 \qquad \text{Divide both sides by 3}$$

$$x = \frac{5}{2} \cdot \frac{1}{3} = \frac{5}{6}$$

$$S = \left\{\frac{5}{6}\right\}$$

(d)   Find S:

$.03x = 1.2$ For this problem, multiply each side by 100 so as to make each of the coefficients integers.

$$100(.03x) = 100(1.2)$$

$$3x = 120$$

$$x = \frac{120}{3}$$

$$x = 40$$

$$S = \{40\}$$

Exercise 21

Find S by using the multiplication of equals axiom:

1. $6x = 18$
2. $5x = 20$
3. $2x = {}^-38$
4. ${}^-3x = 21$
5. ${}^-6x = {}^-24$
6. $\dfrac{3}{5x} = 5$
7. $\dfrac{1}{2}x = 9$
8. $\dfrac{{}^-2}{3}x = \dfrac{1}{4}$
9. $15x = \dfrac{1}{2}$
10. $11x = \dfrac{3}{16}$

11. What is the reciprocal of ${}^-\left(\dfrac{3}{4}\right)$?

12. What is the multiplicative inverse of ${}^-\left(\dfrac{3}{4}\right)$?
13. Name this axiom: $a + 0 = a$.
14. *Find S:* $24x = 96$
15. *Find S:* $3x = 0$
16. If $x = {}^-1$ and $y = {}^-2$, evaluate $4xy - 2y^2 + 3x^2 - 5x^2y$.
17. *Find S:* $x - 8 = 16$
18. *Find S:* ${}^-8x = 16$
19. *Find S:* $3x - 4(x + 2) = 2x + 1$
20. *Find S:* $.003x = .24$
21. List three irrational numbers.
22. What is $\{a, b, c\} \cap \{c, d, e\}$?
23. Write an expression for the sum of the squares of two numbers.
24. *Find S:* $9x = 8x - 12$
25. *Find S:* $\dfrac{3}{7}x = \dfrac{1}{18}$

26. *Find S:* (a) $.01x = .6$
      (b) $3.4x = 102$
      (c) $.006x = 6.024$

4-6. MORE LINEAR EQUATIONS

Now that both axioms for equality have been used, let us put both of them to work in the same equation.

examples:

(a) Find S:

$$3x - 2 = 4$$

$$3x - 2 + 2 = 4 + 2 \qquad \text{Add 2 to each side}$$

$$3x = 6$$

$$\frac{3x}{3} = \frac{6}{3} \qquad \text{Divide each side by 3}$$

$$x = 2$$

$$S = \{2\}$$

(b) Find S:

$$\frac{1}{2}x + 7 = 5$$

$$\frac{1}{2}x + 7 - 7 = 5 - 7 \qquad \text{Subtract 7 from each side}$$

$$\frac{1}{2}x = {}^{-}2$$

$$2\left(\frac{1}{2}x\right) = 2({}^{-}2) \qquad \text{Multiply each side by 2}$$

$$x = {}^{-}4$$

$$S = \{{}^{-}4\}$$

Note: The above examples illustrate the usual procedure of applying the addition axiom *before* the multiplication axiom.

(c) Find S:

$$3(x + 4) - 2(x - 8) + 5x = 3x - 2(x + 1)$$

$$3x + 12 - 2x + 16 + 5x = 3x - 2x - 2$$

$$6x + 28 = x - 2 \qquad \text{Combine similar terms on each side}$$

$$6x + 28 - 28 = x - 2 - 28 \qquad \text{Subtract 28 from each side}$$

$$6x = x - 30$$

$$6x - x = x - x - 30 \qquad \text{Subtract } x \text{ from each side}$$

$$5x = {}^{-}30$$

$$\frac{5x}{5} = \frac{{}^{-}30}{5} \qquad \text{Divide each side by 5}$$

$$x = {}^-6$$
$$S = \{{}^-6\}$$

In all of the work on equations, we have assumed that our solutions were correct. To check this "correctness" proceed as follows:
In Example (a) above we found that $S = \{2\}$.

Check: If $x = 2$, then the *original* equation

$$3x - 2 = 4 \quad \text{becomes}$$
$$3(2) - 2 = 4$$
$$6 - 2 = 4$$
$$4 = 4$$

This last equation is obviously true and the check is complete.
To check example (c) where $S = \{{}^-6\}$.

$$3(x + 4) - 2(x - 8) + 5x = 3x - 2(x + 1) \quad \text{becomes}$$
$$3({}^-6 + 4) - 2({}^-6 - 8) + 5({}^-6) = 3({}^-6) - 2({}^-6 + 1)$$
$$3({}^-2) - 2({}^-14) + ({}^-30) = {}^-18 - 2({}^-5)$$
$${}^-6 + 28 - 30 = {}^-18 + 10$$
$$22 - 30 = {}^-8$$
$${}^-8 = {}^-8$$

This last equation confirms that our solution set is correct.
It is very important that the above procedure for checking be followed. That is, substitute the value for x into each side of the equation, then evaluate each side *independently* of the other. Finally, compare the value of each side of the equation; if each number is the same, then the solution set is correct.

Exercise 22

Find S and Check:

1. $4a - 30 = 2a$
2. $6x + 5 = 7$
3. $a + 4 = {}^-a$
4. $3a + 6 = a + 2$
5. $6x + 7 = 11x + 22$
6. $3x - 4 = 7x + 8$
7. $5m + 2 = 3m + 4$
8. $9x + 3 = 5x - 7$
9. $2m - 7 = 8 - 4m$
10. $3k + 7 = k - 7$

Simplify:

11. $3x(x^2 + x - 2) + 3x - 4(x^2 + 2)$
12. $^-(^-3) - (^-2)(4) + (^-8 - ^-7)$
13. $5a^2 - a + ^-3a + 2a^2 - 7a^2 + 3a - a$
14. $^-6(^-5) + 14 \div (^-7) - ^-8 + 3 - 4$
15. State the distributive axiom.

Find S, if no other question is asked:

16. $3(x - 4) + 2 = 8$
17. $5(x + 2) - 6(x + 7) = 11$
18. $7 = 5x + 2 + 3(2x - 2)$
19. $3(x - 7) = 2 - 2(2x + 1)$
20. $5(3 + 2k) - 3(7 - k) + 6 = 0$
21. Give an example of a second-degree polynomial in two variables.
22. Is the negative of a number always a negative number?
23. $3.4x + 1.2x + 3.4 = ^-5.8$
24. $3.5x = 6.2x - 3.7x + 8$
25. $x^2 - x - 4 = x(x - 2)$
26. State the associative axiom for multiplication.
27. How many members are there in $\{0\}$?
28. $5x(2x - 5) = 10x^2 - 50$
29. $3(2x^2 + x) = 6(4 + x^2) = 0$
30. Is $S = \{3\}$ the solution set for $3x - 4 = 2x + 7$? Show why or why not.
31. $\dfrac{3}{5}x = \dfrac{2}{3}$
32. $x + \dfrac{3}{5} = \dfrac{2}{3}$
33. $3m - 2(m - 6) - 7 = m + 4(2m + 1)$

For problems 34 through 43, find S and check each solution:

34. $^-x = ^-2$ 35. $^-x = 2$

36. $^-2x = ^-2$ 37. $^-x = \dfrac{1}{2}$

38. $^-x - 1 = 2$ 39. $^-5x = \dfrac{1}{5}$

40. $^-(^-x) = {}^-\left(\dfrac{1}{2}\right)$

41. $^-x + \dfrac{1}{2} = \dfrac{1}{2}$

42. $\dfrac{1}{3}x - \dfrac{1}{3} = \dfrac{1}{3}$

43. $.6x = .003$

44. List the members of $\{x \mid x$ is a positive even integer less than 12$\}$.

45. Give a rational number that is between 6 and 7.

46. Draw a number line and place on it the real numbers $^-6$, $^-1/3$, 2, $^-15/7$, 5, and $^-\sqrt{2}$.

47. Give an example of a negative irrational number.

48. Write an algebraic expression for "three less than twice some number."

49. State the multiplication of equals axiom.

50. Find S if $8(2 - y) + 7(3 - y) + 7 = 9(y + 2) - 3(y + 4)$, and check the solution.

4-7. FORMULAS AND LITERAL EQUATIONS

The application of linear equations in one variable to scientific work is quite widespread. Most of the common elementary formulas of chemistry, physics, and geometry are linear equations. A formula such as $A = lw$ may be given. However, in an applied situation the values for A and w may be known. Because in this situation we want to find the value for l, it is better to have a formula in which l is by itself on one side of the equation.

The formula $A = lw$ could be written

$l = \dfrac{A}{w}$ by dividing each side by w

examples: (a) Solve for m:

$$a = m + b$$
$$a - b = m + b - b$$
$$a - b = m$$
$$m = a - b \quad \text{using the symmetric axiom}$$

(b) Solve for t^2

$$s = \frac{1}{2}gt^2$$
$$2s = gt^2$$

$$\frac{2s}{g} = t^2$$

$$t^2 = \frac{2s}{g}$$

(c)   Solve for b_2

$$A = \frac{1}{2}h(b_1 + b_2)$$

$$2A = h(b_1 + b_2) \qquad \text{Multiply each side by 2}$$

$$\frac{2A}{h} = b_1 + b_2 \qquad \text{Divide each side by } h$$

$$\frac{2A}{h} - b_1 = b_2$$

$$b_2 = \frac{2A}{h} - b_1$$

Exercise 23

Solve for the specified variable. Do not use solution sets:

1. $P = 4s$, for s
2. $A = lw$, for w
3. $C = 2\pi r$, for r
4. $A = \frac{1}{2}h(b_1 + b_2)$
5. $V = v_1 + v_2$, for v_1
6. $A = P + Prt$, for t
7. $V = lwh$, for h
8. $V = \frac{4}{3}\pi r^3$, for r^3
9. $S = 4\pi r^2$, for r^2
10. $A = \frac{1}{2}h(b_1 + b_2)$, for b_1
11. $C = \frac{5}{9}(F - 32)$, for F
12. $K = gt + t_0$, for t
13. $P = a + b + c$, for a
14. $V = \frac{KT}{P}$, for P
15. $Q = mv$, for v
16. The volume of a rectangular solid is given by the formula $V = lwh$, where l is length, w is width, and h is height. What is the length of a solid whose volume is 256 cubic inches if the width is 16 inches and the height is 8 inches?

17. In the formula $A = P + Prt$, A is the amount of money paid back, P is the amount of money borrowed, r is the yearly rate (%) of interest, and t is the time in years. If Joe paid back \$318 on a 6% loan which he held for one year, how much did he borrow?

18. What is the length of a rectangle if its perimeter is 160 feet and its width is 20 feet? (Use $P = 2l + 2w$.)

Find S:

19. $2x = 13$

20. $x + 8 = {}^-8$

21. $3x + 4 = 5$

22. $18 - a = 2a - 3$

23. $3(a + 2) = {}^-9$

24. $5a - 6 = 12a + 7$

25. $2x - \dfrac{1}{2} = \dfrac{1}{5}$

26. $3(x + 2) = 2(x + 3)$

27. $5 = 8$ (Think carefully.)

28. $5x = 8x$

4-8.   STATED PROBLEMS

One of the objectives of studying algebra is the solution of specific applied problems. In practice, problems will be proposed verbally or in written form using words rather than symbols. We must study what is being stated in the problem and translate it into an equation. The thinking that is required for this translation is an end in itself. Do not be surprised if the problems that follow seem to have no application as far as you are concerned.

Sample translations:

(a)   Two consecutive integers:
  let n = first integer and $(n + 1)$ the next integer

(b)   Two consecutive even integers:
  let n = first even integer and $(n + 2)$ the next even integer

(c)   Two consecutive odd integers:
  let n = first odd integer and $(n + 2)$ the next odd integer

(d)   If a total of \$10 is to be spent for shirts and ties, use x dollars for shirts and $(10 - x)$ dollars for ties.

(e)   If a man has 6 nickels, then he has 5(6) cents.
  If a man has k nickels, then he has $5k$ cents.

(f)   If a man has d dimes and q quarters, then he has $(10d + 25q)$ cents total.

(g) If a man deposits m dollars at 5% interest, then he receives 5% of m dollars in interest, that is, interest = $.05m$.

examples: (a)   Find three consecutive integers whose sum is 96.

Solution:

Let n, $(n + 1)$, $(n + 2)$ represent the three consecutive integers.

The sum of these integers is

$$n + (n + 1) + (n + 2) = 96$$
$$3n + 3 = 96$$
$$3n = 93$$
$$n = 31$$
$$n + 1 = 32$$
$$n + 2 = 33$$

Answer: The three integers are 31, 32, 33.
Check: $31 + 32 + 33 = 96$

(b) If three times a certain number is increased by 7, the result is the same as subtracting twice the number from 32. Find the number.

Solution:

Let x = the number

Then $3x + 7 = 32 - 2x$ is the equation described.

$$3x + 7 = 32 - 2x$$
$$5x = 25$$
$$x = 5$$

Answer: 5 is the number
Check: $3 \cdot 5 + 7 = 32 - 2 \cdot 5$
$$15 + 7 = 32 - 10$$
$$22 = 22$$

(c) Sam has a collection of nickels and dimes consisting of 42 coins. The coins are worth a total of $3.45. How many of each kind are there?

Solution:

Let
$$x = \text{number of nickels}$$
$$42 - x = \text{number of dimes}$$
$$\text{Penny value of nickels} = 5x \text{ (cents)}$$

Penny value of dimes $= 10(42 - x)$ cents

Value of nickels + value of dimes = total value of collection

$$5x + 10(42 - x) = 345$$

Note: All terms in this equation are given in pennies. An *incorrect* equation $x + (42 - x) = 345$ does not have consistent labels, since the total numbers of coins, $x + (42 - x)$, is related to a value, 345, in cents.

Now solve the penny value equation.

$$5x + 10(42 - x) = 345$$
$$5x + 420 - 10x = 345$$
$$^{-}5x = ^{-}75$$
$$x = 15 \text{ nickels}$$
$$42 - x = 42 - 15 = 27 \text{ dimes}$$

Answer: 15 nickels and 27 dimes in the collection

Check: 15 nickels are worth $15(5¢) = 75¢$
27 dimes are worth $27(10¢) = 270¢$
$75¢ + 270¢ = 345¢ = \$3.45$

(d) Mr. Ruskauff invested some money in stocks that paid 7% yearly interest. He invested twice as much money in bonds that paid 5% yearly interest. At the end of one year, he had earned \$34 total interest. How much did he invest in each place?

Solution:

Let

$$x = \text{dollars invested at 7\%}$$
$$2x = \text{dollars invested at 5\%}$$
$$7\% \text{ of } x = .07x = \text{interest received from 7\% stocks}$$
$$5\% \text{ of } 2x = .05(2x) = \text{interest received from 5\% bonds}$$

Interest from stocks + interest from bonds $= \$34$

$$.07x + .05(2x) = 34$$
$$7x + 5(2x) = 3400 \quad \text{Multiply each side by 100}$$
$$17x = 3400$$
$$x = 3400 \div 17$$
$$x = \$200$$
$$2x = \$400$$

Answer: $200 invested at 7% interest (stocks)
           $400 invested at 5% interest (bonds)

Check: 7% of $200 is $14 and 5% of $400 is $20
          Total interest is $14 + $20 = $34

Exercise 24

Write an algebraic expression for each statement:

1. One number is 3 times as large as another number:
   - (a) Represent each number if x is the smaller number.
   - (b) Represent 4 more than the larger number.
   - (c) Represent the sum of the two numbers.
   - (d) Represent twice the smaller plus the larger.

2. Three numbers are *consecutive even* integers:
   - (a) Represent the numbers if the smallest is k.
   - (b) Represent their sum.
   - (c) Represent the first plus twice the second minus 3 more than the third.
   - (d) If the middle number is x, represent the three numbers.

3. Given n nickels, d dimes, and p pennies:
   - (a) Represent the total number of coins.
   - (b) Represent the penny value of the nickels.
   - (c) Represent the total penny value of the nickels and dimes.
   - (d) Represent the total penny value of all the coins.

4. A board is 40 feet long:
   - (a) Cut the board into two pieces and represent each piece. Let m represent one piece.
   - (b) The sum of the two pieces should be 40 feet. What is the sum of your two pieces?

5. Represent the penny value of k quarters.

6. Represent a certain number added to negative ten.

7. Represent a certain number less 7.

8. Represent 7 less than a certain number.

9. Represent 7 diminished by a certain number.

10. Represent the yearly interest received on m dollars at 4½%.

11. Represent the number of passing students in a class of 75% passed.

12. Represent the number of feet in the smaller portion of a 50-foot piece of rope if the larger portion is *y* feet long.

13. A collection of coins contains twice as many nickels as dimes:
    (a) How many of each coin are there?
    (b) What is the penny value of the nickels?
    (c) What is the penny value of the dimes?
    (d) What is the total penny value?

14. If $(a + b)$ dollars is invested at 12% interest, what is the amount of interest?

For each of the following problems:
    (a) *choose a variable and state in words what it represents;*
    (b) *represent other quantities in terms of this variable;*
    (c) *write an equation in terms of this variable;*
    (d) *solve the equation;*
    (e) *give all required answers;*
    (f) *check answers with the stated problem.*

15. Find two consecutive integers whose sum is 53.

16. Find three consecutive integers whose sum is 60.

17. Find three consecutive *odd* integers whose sum is 93.

18. Find a number such that 4 more than twice the number equals 3 times the number.

19. If a 24-foot board is cut into two parts so that one part is 8 feet longer than the other, how long is each part?

20. A man has $210 to give to his wife and son. He gives his wife twice as much as to his son. How much did each receive?

21. A fence is placed around a rectangular area. If the length is 80 feet more than the width and the roll of fencing is 500 feet long, what are the dimensions of the area? [Make a sketch and use the formula Perimeter = 2(length + width).]

22. What is the length of a play yard if the perimeter is 120 feet and the width is 2 feet less than the length?

23. A truck and camper together cost $7486. If the truck costs $2124 more than the camper, what is the cost of each?

24. One number is 6 more than another. If 3 times the smaller number is added to the larger number, the result is the same as 8 more than twice the larger. Find the numbers.

25. Jerry has 8 more nickels than dimes. He has $1.15 total. How many nickels does he have?

26. Tess bought some 9¢ stamps and twice as many 12¢ stamps. She gave the clerk a ten-dollar bill and received $5.05 in change. How many of each kind of stamp did she get?

27. Two numbers have a sum of ⁻16. One number is 3 more than 5 times the other. What is the smaller number?

28. There are 18 coins on the table. Some are dimes and the rest are quarters. The value of these coins is $2.85. What is the *value* of the dimes?

29. For a concert 600 tickets were sold. Adults' tickets cost $2.50 each and children's tickets cost $1.50 each. A total of $1120 was received at the box office. How many children had tickets? Assuming all adult tickets were sold to couples only, how many adult couples attended?

30. On kid's day at the baseball park the children outnumbered the adults five to one. If adults paid $3 each and children paid $1 each, how many tickets were sold if $1120 was taken in?

31. The Weber family saved $1200 one year. Some went into a bank savings account that paid 4% interest and the rest into a credit union account that paid 5½%. The total interest earned was $60. How much of the original was in the credit union?

32. A man received the same amount of interest from two investments. One paid 7% and the other 5% yearly. The total amount invested was $600. What was the total interest received?

33. Three numbers are given. The first is twice the third and the second is 4 more than the first. Their sum is 39. What are the numbers?

34. Steve has 80 coins consisting of pennies, nickels, and dimes. He has 8 more pennies than nickels. The value of all the money is $6.44. How many dimes are there?

35. John is now 6 years older than Anne. In 2 years John will be twice as old as Anne. How old is each now?

36. At present Mrs. West is 3 times as old as her son. In 12 years she will be twice as old as her son. How old was each 5 years ago?

37. If 300 pounds is 28% of the weight of a steer, then what is the weight of the steer?

38. Divide the number 55 into two parts so that twice one part is 5 more than the other.

39. A total of $700 is invested. Some earns 6% interest and the remainder earns 8% interest. The interest from the 8% investment is $3 greater than that from the 6% investment. How much was placed in each account?

40. To stock a pantry, the following purchases were made: several 65¢ cans of meat, twice as many 30¢ cans of beans as meat, 3 more cans of 40¢ fruit juice than cans of meat, and 2 less cans of 20¢ soup than cans of beans. A 20-dollar bill was given to the clerk to pay for these items and $8.95 in change was received. How many cans of soup were purchased?

Cumulative Review: Chapters 1 through 4

1. State three axioms from memory.
2. What is another name for linear?
3. The sign of the product of an odd number of negative factors is _____ .
4. If two sets contain no common members, then
   (a) the intersection of the sets contains _____ .
   (b) the union of the sets contains _____ .
5. What is a solution set, S?
6. *Simplify:* $^-8 + 3 - (4 + ^-2) - 6 - ^-3 + 1 + ^-4$
7. *Simplify:* $3a^2 + 2a + 6 - (4a^2 + 5a - 7) + (7a^2 - 2a - 8)$
8. *Multiply and simplify:* $(3x^2 + 2x + 5)(x^2 - 8x - 1)$
9. If $x = ^-3$ and $y = 4$, evaluate $2x^2 - 3xy - y^2$.
10. Find three consecutive even integers whose sum is 2.
11. Include $3x^2$, ^-4x, and $^-5$ in parentheses preceded by a negative sign.
12. Give an example of a quadratic trinomial in one variable.
13. (a) $^-8(^-4) = ?$ (b) $^-8 \div ^-4 = ?$ (c) $^-8 - ^-4 = ?$
14. *Find S:* $3x - 2 = 2(x + 3) - 7$
15. The set that contains all of the integers, rational numbers, and irrational numbers is called the set of _____ _____.

5 Linear inequalities*

5-1. THE CONCEPT OF INEQUALITIES

In ordinary conversation and reading it is quite common to compare quantities that are not equal. For example: "Paul weighs more than Tim"; or "I make less money than Sally."

To express these ideas mathematically, symbols are used. The statement "6 is less than 8" is written $6 < 8$. Conversely, "8 is greater than 6" may be written $8 > 6$. To express the idea that children six years old or younger will be admitted free to an event, we may write $x \leq 6$, where x represents all children admitted free. The inequality $x \leq 6$ is translated as "x is less than *or* equal to 6." If $x \leq 6$, then $3 \leq 6$ and $6 \leq 6$ are each true.

5-2. AXIOMS THAT PERTAIN TO INEQUALITIES

Since inequalities are being introduced for the first time, we must have some "rules" of procedure. As in the past, these rules are called axioms.

The idea $a < b$ is accepted as an undefined term. Consequently, we are able to define $a > b$ formally.

DEFINITION: *$a > b$ is equivalent to $b < a$.*

---

*Optional Chapter

To gain some insight into the meaning of the undefined term, $a < b$, we can observe all numbers on the real number line and claim that any number is smaller than a number to its right.

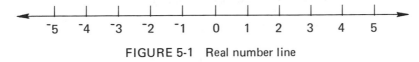

FIGURE 5-1   Real number line

In Figure 5-1 we observe that $3 < 5$ and $^-2 < ^-1$.

AXIOMS FOR INEQUALITIES: *(a, b, c ϵ R)*

1.   *If $a < b$, then $a + c < b + c$.*
2.   *If $a < b$ and $c > 0$, then $ac < bc$.*
3.   *If $a < b$ and $c < 0$, then $ac > bc$.*
4.   *Transitive axiom: If $a < b$ and $b < c$, then $a < c$.*
5.   *Trichotomy axiom: One and only one of the following is always true:*
       (i) $a < b$
       (ii) $a = b$
       (iii) $a > b$

It must be noted that Axiom 3 is the major difference between solving inequalities and equations. This axiom requires that the inequality reverse its direction if we multiply both sides by a negative number. An example will illustrate: Given $3 < 4$, multiply both sides by ($^-1$) and we get $^-3 > ^-4$, which agrees with the position on the number line.

5-3.   THE SOLUTION SET OF A LINEAR INEQUALITY AND ITS GRAPH

In chapter 4 we learned that $x + 3 = 5$ is a linear equation in one variable, whose solution set is $S = \{x \mid x + 3 = 5\} = \{2\}$. The statement $x + 3 < 5$ also has a solution set, $S = \{x \mid x + 3 < 5\}$. However, listing the members of this set is not possible, as it is an infinite set. To give a more precise description of the set $S = \{x \mid x + 3 < 5\}$, let us apply Axiom 1 of section 5-2:

$$\text{If } x + 3 < 5, \text{ then } x + 3 + (^-3) < 5 + (^-3)$$
$$x < 2$$

The inequality is true if x is any real number less than 2, which leads to $S = \{x \mid x < 2\}$.

Since it is not possible to list every member of S, it is sometimes convenient to make a graph (picture) of S.

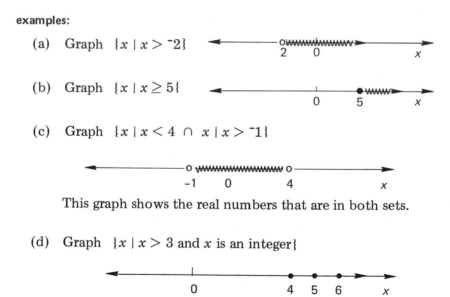

FIGURE 5-2 Graph of $\{x \mid x < 2\}$

Figure 5-2 indicates that any real number to the left of 2 is in S. Note that $2 \notin S$, and this is denoted by the *open dot* at 2. If we graph $\{x \mid x \leq 2\}$, which does contain 2, then a *closed dot* is used at 2.

examples:

(a)   Graph $\{x \mid x > ^-2\}$

(b)   Graph $\{x \mid x \geq 5\}$

(c)   Graph $\{x \mid x < 4 \cap x \mid x > ^-1\}$

This graph shows the real numbers that are in both sets.

(d)   Graph $\{x \mid x > 3$ and x is an integer$\}$

Exercise 25

1.   By definition, what does $8 > 3$ mean?
2.   *True or false:*
    (a) $3(^-4) > 0$ (b) $6 + 7 \leq 13$ (c) $3 + 5 < 4^2$
    (d)   If $S = \{x \mid x \geq 4\}$, then $4.001 \notin S$.
3.   *If possible, list all the members of each set.*
    (a) $A = \{x \mid x$ is a positive integer and $x < 5\}$
    (b) $B = \{x \mid x$ is a positive real number and $x < 5\}$
    (c) $C = \{x \mid x$ is an integer and $x \geq 7\}$
4.   *Referring to problem 3, give the largest and smallest member of each set if possible.*

5. *Refer to problem 3b:*
   (a) Is $5 \in B$?   (b) Is $4.999 \in B$?
   (c) Can you name a number larger than 4.999 which is in B?

6. *Refer to problem 3:*
   (a) Find $A \cap B$ (b) Is A a subset of B?
   (c) Find $A \cap C$

7. *Graph each set on a separate number line:*
   (a) $\{x \mid x < {}^-2\}$
   (b) $\{x \mid x \leq 9\}$
   (c) $\{x \mid x \geq {}^-5\}$
   (d) $\{x \mid x \geq {}^-3\} \cap \{x \mid x < 5\}$
   (e) $\{x \mid x > 1\} \cap \{x \mid x < 2\}$
   (f) $\{x \mid x < {}^-3\} \cap \{x \mid x > {}^-2\}$
   (g) $\{x \mid x < 4$ and x is a positive integer $\}$
   (h) $\{x \mid x \geq {}^-3$ and x is a negative integer$\}$
   (i)   The intersection of (g) and (h) above.
   (j) $\{x \mid x > 4\} \cap \{x \mid x > 3\}$
   (k) $\{x \mid x \leq {}^-6\} \cap \{x \mid x < {}^-3\}$

8. Why must $x = 6$ be false if $x > 7$
   (*Hint:* Two axioms apply to this problem.)

9. *Graph each set on a separate number line:*
   (a) $\{x \mid x \leq 3$ or $x \geq 5\}$
   (b) $\{x \mid x \geq 2$ and $x < 6\}$
   (c) $\{x \mid x < {}^-4$ or $x > 5\}$
   (d) $\{x \mid x < {}^-4$ and $x > 5\}$
   (e) $\{x \mid x \leq 10$ and $x \geq {}^-4$ or $x < {}^-6\}$
   (f) $\{x \mid x > 2$ and $x > 5\}$

5-4. APPLICATION OF THE AXIOMS FOR INEQUALITIES

In section 5-2 we stated five axioms. The first three of these axioms show us how to perform the manipulations that give the solution set for an inequality.

examples: (a)   Find S:

$$x + 7 < 9$$
$$x + 7 + {}^-7 < 9 + {}^-7 \qquad \text{(Axiom 1, section 5-2)}$$
$$x < 2$$

Answer: $S = \{x \mid x < 2\}$

(b)   Find S:

$$3x < 15$$
$$\frac{1}{3} \cdot 3x < \frac{1}{3} \cdot 15 \qquad \text{(Axiom 2, section 5-2)}$$
$$x < 5$$

Answer: $S = \{x \mid x < 5\}$

(c)   Find S:

$${}^-x < 4$$
$${}^-1({}^-x) > {}^-1(4) \qquad \text{(Axiom 3, section 5-2)}$$
$$x > {}^-4$$

Answer: $S = \{x \mid x > {}^-4\}$

(d)   Find S:

$$13x + 8 \le 2x - 9$$
$$11x + 8 \le {}^-9 \qquad \text{(Axiom 1)}$$
$$11x \le {}^-17 \qquad \text{(Axiom 1)}$$
$$x \le {}^-\left(\frac{17}{11}\right) \qquad \text{(Axiom 2)}$$

Answer: $S = \left\{x \mid x \le {}^-\left(\frac{17}{11}\right)\right\}$

(e)   Find S:

$${}^-3x + 2({}^-x + 6) \ge 3x - 4({}^-3x - 7)$$
$${}^-3x - 2x + 12 \ge 3x + 12x + 28$$
$${}^-5x + 12 \ge 15x + 28$$
$${}^-20x + 12 \ge 28$$
$${}^-20x \ge 16$$
$$x \le {}^-\left(\frac{16}{20}\right)$$
$$x \le {}^-\left(\frac{4}{5}\right)$$

Answer: $S = \left\{x \mid x \le {}^-\left(\frac{4}{5}\right)\right\}$

Exercise 26

Find S:

1. $x + 6 < 3$
2. $x + 6 = 3$
3. $2x - 3 > 4$
4. $5x - 5 < 12$
5. $2x + 1 \leq x - 3$
6. $x + {}^-3 \geq {}^-6$
7. ${}^-x \leq {}^-5$
8. ${}^-2x \leq 12$
9. $x + 4 > 5x + 8$
10. $5 - 8x < 2x + 15$
11. $3(x + 2) < x + 1$
12. ${}^-x - 8 = x + 4$
13. $2(x + 5) - 5(x + 2) < 0$
14. ${}^-3x + 7(4 - x) < 2x + 1$
15. $3x(x + 2) + 5 < x^2 + 2x(x - 7)$
16. State the trichotomy axiom.
17. *Graph:* $\{x \mid x \geq {}^-3\}$
18. *Graph:* $\{x \mid 2x - 7 < 5x + 4\}$
19. *Graph:* $\{x \mid 4x + 1 > 2\} \cap \{x \mid x - 5 < 2\}$
20. *Graph:* $\{x \mid x \leq 3\} \cap \{x \mid {}^-24x \leq {}^-72\}$

6
Special products and factoring

6-1. WHAT IS FACTORING?

By definition, a factor is a number that is multiplied. *The process of finding factors of a number or an expression is called factoring.* All numbers have trivial factors, 1 and the number itself. If a number has factors other than 1 and itself it is called a *composite number.* A number larger than 1 with no factors except 1 and itself is called a *prime number.* The factors of a number which are themselves prime numbers are called the *prime factors* of that number.

examples:

(a) 2, 3, 5, 7, 11, 13, 17, 19, 23, and 29 are the *prime numbers* between zero and thirty.

(b) 6 is a *composite number*, since 6 = 3 · 2

(c) 3 and 2 are the *prime factors* of 6.

(d) 2, 2, 2, and 2 are the *prime factors* of 16.

The ability to find prime factors depends upon a knowledge of the multiplication tables and certain divisibility tests. For example, all even numbers are divisible by 2 and have a factor of 2. The number 3 is a factor of a number if the sum of the digits in the number is divisible by 3; for example, 51 is divisible by 3 since 5 + 1 = 6 and 6 is divisible by 3.

examples:

    (a)   Find the prime factors of 432.

          To begin, we note that 432 is an even number and can be divided by 2.

$$2\overline{)432} = 216 \qquad \text{which means} \quad 432 = 2 \cdot 216$$

$$2\overline{)216} = 108 \qquad\qquad 432 = 2 \cdot 2 \cdot 108$$

$$2\overline{)108} = 54 \qquad\qquad 432 = 2 \cdot 2 \cdot 2 \cdot 54$$

          At this point the times tables tell us that $54 = 9 \cdot 6$ giving

$$432 = 2 \cdot 2 \cdot 2 \cdot 9 \cdot 6$$
$$= 2 \cdot 2 \cdot 2 \cdot 3 \cdot 3 \cdot 3 \cdot 2$$
$$= 2^4 \cdot 3^3 \text{ as the seven prime factors of } 432.$$

    (b)   Find the prime factors of 1365.

          Since 1365 ends in 5, it is divisible by 5.

$$5\overline{)1365} = 273 \qquad \text{which means} \quad 1365 = 5 \cdot 273$$

          The number 273 is divisible by 3, as $2 + 7 + 3 = 12$ which is divisible by 3.

$$3\overline{)273} = 91 \qquad \text{So far we have } 1365 = 5 \cdot 3 \cdot 91$$

          To factor 91, trial and error produces 7 as a factor.

$$7\overline{)91} = 13 \qquad \text{Finally, } 1365 = 5 \cdot 3 \cdot 7 \cdot 13$$

Exercise 27

Find the prime factors of each number:

1. 10	6. ⁻6	11. 88	16. 64
2. 20	7. ⁻8	12. 51	17. 98
3. 45	8. ⁻40	13. 78	18. 500
4. 100	9. 125	14. 160	19. 750
5. 48	10. 210	15. 1000	20. 1728

Find S:

21. $4x - 3 = 10x$ 23. $2(x - 3) + x = 5(x + 1)$
22. $5x = \frac{1}{2}$ 24. $3(2 - x) - x = 0$
25. State the distributive axiom.
26. Explain why 217,242 is divisible by 3.

6-2.   THE DISTRIBUTIVE AXIOM

We will build a "multiplication table" for algebraic expressions by special products. These special products must be associated with their factors.

The most common multiplication in algebra involves the distributive axiom, $a(b + c) = ab + ac$. The left side of the equation contains two factors, the monomial a and the binomial $(b + c)$. The right side, $ab + ac$, contains only one factor but two terms. The important observation here is that each of these two terms contains a common factor, a. The critical part of Exercise 28 is to relate the product obtained with its factors.

examples:

(a) $6(3a + 5b) = 18a + 30b$

The 6 is "distributed" over each term inside the parentheses. Accordingly, each term in the right-hand product contains this common factor, 6.

(b)   The distributive axiom involving variables can be used when values are assigned to a, b, and c.

That is, $6(83) = 6(80 + 3)$
                    $= 6 \cdot 80 + 6 \cdot 3$
                    $= 480 + 18$
                    $= 498$

This method of multiplication is convenient when multiplication has to be done mentally.

Exercise 28

Apply the distributive axiom in each problem and name the common factor that is introduced into each term of the product, unless otherwise specified:

1. $3(a + b)$
2. $5(2a + 1)$
3. $x(2a + b)$
4. $a(a + 1)$
5. $2(p^2 + 4)$
6. $ax(b + c)$
7. $^-3(a + 5)$
8. $^-m(m + 2)$

9. $3a(a + 5)$
10. $t(t + 1)$
11. $^-k(k - 2)$
12. $a^2(x + 4)$
13. $4m(m + 3)$
14. $xy^2(a + b + c)$
15. $2m(m + n - 3)$
16. $3x(6a + 3b + 5)$

17. *Find S:* $3(x + 4) - 1 = {}^-4$
18. *Find S:* $\frac{3}{4}x = 5$
19. What is the multiplicative identity?
20. How many factors are there in the expression $3(x + 2a)(xy + 3)$?
21. What does the addition axiom for equality allow us to do?
22. $m^2(2a + b)$
23. $ax(y + z + 2)$
24. $a(a^3 + 2a^2 + 3a + 2)$
25. $6k(2k^2 + 3k + 1)$
26. $a^2(x^2 + y^2 + z^2)$
27. Reread section 1-1 and give the necessary parts of a mathematical system.
28. $^-3a(a^3 - 2a^2 + 3a - 1)$
29. $abx(a^2 + ab + 3)$
30. $mn(3m^2 - 4mn + 2n^2)$
31. $pq^2(p^3 + 3p^2q - 4pq^2 - 6q^3)$
32. If $A = \{1, 3, 5, 7, 9\}$ and $B = \{2, 4, 6, 8\}$ form
    (a) $A \cap B$ (b) $A \cup B$
33. Give examples of two rational and two irrational numbers.

For Problems 34 to 44, use the distributive axiom as shown in example (b), page 70:

34. $5(42)$
36. $9(56)$
38. $3(49)$

35. $7(24)$
37. $4(79)$
    (Use $4(80 - 1)$)

39. 6(36)   40. 2(123)
41. 8(96)   42. 4(99)
43. 6(28)   44. 9(244)

6-3.   REMOVING A COMMON FACTOR

In section 6-2 we very carefully reviewed the distributive axiom to observe how this axiom distributed a common factor over two or more terms. Now we shall reverse that process by removing the common factor from two or more terms in an algebraic expression. The symmetric axiom for equality (if $a = b$ then $b = a$) allows us to begin with either side of an equation. Apply this axiom to the distributive axiom and we get "if $a(b + c) = ab + ac$ then $ab + ac = a(b + c)$." The last clause gives a pattern for "removing a common factor, a."

examples: Remove the common factor:

(a) $3a + 3b = 3(a + b)$
(b) $mx + my = m(x + y)$
(c) $\bar{}ax - ay = \bar{}a(x + y)$ or $a(\bar{}x - y)$
(d) $pq^2 - pq - p = p(q^2 - q - 1)$
(e) $m^3 - 3m^2 + 2m = m(m^2 - 3m + 2)$
(f) $21m^2 - 30m = 3m(7m - 10)$

In example (f) it may be better to put in an intermediate step.
$$21m^2 - 30m = (3 \cdot 7m \cdot m) - (3 \cdot 10m)$$
$$= 3m(7m - 10)$$

By factoring each term, the common factor is easier to identify.
Also, notice that in example (d) the factor q is not removed, as it is not common to all three terms.

Exercise 29

Remove the largest common factor from each algebraic expression, if there is a common factor:

1. $2a + 2b$ 3. $5a - 10b$
2. $6a + 12b$ 4. $16y + 12$

5. $11m - 33$
6. $18k - 24$
7. $5x + 10xy$
8. $12ab + 2$
9. $3a + 16$
10. $3a + 2ab$
11. $9m^2 - 12m$
12. $3a^2 x^2 + 5a^2 y$
13. *Multiply:* $5x^2 (5x + 2)$
14. *Simplify:* $2a(a + 3) - 4(a^2 + 2a + 3)$
15. *Factor:* $15k^2 - 3k$
16. *Find S:* $3(x - 2) - 4(2 - x) = x + 1$
17. Explain "factoring."
18. *Factor:* $9a^2 b - 25x^2 y$

Remove the common factor:

19. $ax^2 - ay^2$
20. $3a^2 - 6a + 3$
21. $8t^2 - 2$
22. $30x^3 - 20x^2 - 10x - 10$
23. $pqr - qrt - pqt - prt$
24. $mnp + mpq - mnq$
25. $16x^3 + 20x^2 - 14x - 12$
26. $20x^2 y^2 + 24x^3 y^2 - 16x^2 y$
27. $R^2 - r^2$
28. $\frac{1}{2} mv_1 - \frac{1}{2} mv_2$
29. $P + Prt$
30. $fk_1 - fk_2 - f^2$

Solve each equation for the indicated variable:

example: $K = ab + ac$, for a
$$K = ab + ac$$
$$K = a(b + c)$$
$$\frac{K}{b + c} = a \quad \text{or} \quad a = \frac{K}{b + c}$$

31. $A = P + Prt$, for P
32. $F = 2kc_1 - kc_2$, for k
33. $V = gh + gh$, for h
34. $V = \pi r^3 + \pi R^3$, for R^3
35. $S = 2mv_1 + 3mv_2 + 2mv_3$, for m
36. $Qm = mk - 3$, for m
37. $T_2 = 2gh_1 + 2gh_2$, for g
38. $D = Kx_1 - Kx_2$, for K
39. $D = Kx_1 - Kx_2$, for x_2
40. *True or false:* A factor must appear in all three terms of a trinomial if it is to be removed as a common factor.

6-4. THE PRODUCT OF THE SUM AND DIFFERENCE OF TWO TERMS

The binomials $(a + b)$ and $(a - b)$ contain the same two terms. The first binomial contains the sum of a and b while the second binomial contains the difference when b is subtracted from a. The product is given by the following theorem.

THEOREM 6-1: $(a + b)(a - b) = a^2 - b^2$

Proof:

$$
\begin{aligned}
(a + b)(a - b) &= (a + b)a - (a + b)b && \text{distributive axiom and} \\
&&& \text{definition of subtraction} \\
&= a(a + b) - b(a + b) && \text{commutative axiom for} \\
&&& \text{multiplication} \\
&= a^2 + ab - ba - b^2 && \text{distributive axiom} \\
&= a^2 - b^2 && \text{additive inverse axiom}
\end{aligned}
$$

The formula in the above theorem is the second "special product" to be considered. Again it is important to associate the factors with their product. In this case the product is called the *difference of two squares*. Notice that the two terms are in the same order in each binomial and that the same order appears in the product.

examples:

(a) $(a + 3)(a - 3) = a^2 - 9$

(b) $(xy + 2m)(xy - 2m) = x^2 y^2 - 4m^2$

(c) $32 \cdot 28 = (30 + 2)(30 - 2) = (30)^2 - 2^2 = 900 - 4 = 896$

(d) $108 \cdot 92 = (100 + 8)(100 - 8) = (100)^2 - 8^2 = 10,000 - 64 = 9936$

Exercise 30

Perform the indicated operations:

1. $(x + 5)(x - 5)$ 6. $(2m - 3)(2m + 3)$
2. $(x + 7)(x - 7)$ 7. $(6a + 5)(6a - 5)$
3. $(ab + 1)(ab - 1)$ 8. $(x - 40)(x + 40)$
4. $(2a + 1)(2a - 1)$ 9. $(pqr - 3)(pqr + 3)$
5. $(20 + 1)(20 - 1)$ 10. $(7x^2 + y)(7x^2 - y)$
11. Is $x^2 - y^2$ a perfect square? Why?
12. *Factor:* $3x - 6$ 13. $2a(3a + 4b)$
14. $(8a + 2)(8a - 2)$ 15. $5(x + 3) - 2(x + 7)$
16. *Factor:* $axy - 2axz - 3axw$
17. *Factor:* $pq^2 - pq$
18. *Find S:* $3a - 6a + 2(a + 1) = 0$

19. Find two consecutive odd integers whose sum is 68.
    (*Remember:* You must always write an equation and solve it.)
20. If Bob has twice as many nickels as dimes and he has 80¢ total,
    how many coins does he have?
21. $(3x + 1)(3x - 1)$ 22. $(a + 5x)(a - 5x)$
23. $(2mn + 11)(2mn - 11)$ 24. *Solve for g:* $V = gh + gm$
25. *Solve for P:* $A = P + Prt$
26. *Solve for m:* $3m + 1 = am - 5$
27. *Solve for y:* $y = 2x + yk$
28. *Solve for x:* $Q = Kx + pz - 2x$

Perform the multiplication as in examples (c) and (d), page 74:

29. $41 \cdot 39$ 30. $26 \cdot 34$
31. $53 \cdot 47$ 32. $106 \cdot 94$
33. $208 \cdot 192$ 34. $75 \cdot 85$
35. $99 \cdot 101$

6-5. FACTORING THE DIFFERENCE OF TWO SQUARES

We should see that each time the sum and difference of two terms
is multiplied a *difference of two squares* is the product. To factor the
difference of two squares, all that is necessary is to determine which
two numbers are squared, and then read the theorem of section 6-4
from right to left, that is, $a^2 - b^2 = (a + b)(a - b)$.

examples: (a) $x^2 - 16 = x^2 - (4)^2 = (x + 4)(x - 4)$
                  (b) $9x^2 - 25 = (3x)^2 - (5)^2 = (3x + 5)(3x - 5)$

   However,

                  (c) $x^2 + 4$ is prime and cannot be factored when
                        treating real numbers only.
                  (d) $6x^2 - 150 = 6(x^2 - 25) = 6(x + 5)(x - 5)$
                        (It is not necessary to write 6 as $3 \cdot 2$.)

In example (d) we see an example of two types of factoring. In all
cases we should look for a common factor, remove it, and then exam-
ine the remaining polynomial for further factorization.

Note: To "factor completely" means to separate an expression
into its prime factors. *All monomials will be considered prime.*

Exercise 31

Factor completely, unless otherwise stated:

1. $x^2 - 36$
2. $x^2 - 49$
3. $9x^2 - 1$
4. $25a^2 - 1$
5. $a^2 - 1$
6. $5x^2 - 20$
7. $8x^2 - 2$
8. $4x^2 - 16$
9. $4x^2 - 25$
10. $15x^2 - 60$
11. $3x^2 - 9$
12. $6x^2 - 36$
13. $4x^2 - 49$
14. $ax^2 - a$
15. $ax^3 - ax$
16. $ax^3 - ax^2$

17. State the commutative axiom for addition.
18. How many factors are there in each of the following expressions?
    (a) $ax^2 - bx + c$ (b) $3(ax + b)$
    (c) $(a + b)(c + d)$
19. What is the numerical coefficient in the monomial $18x^3$?
20. *Factor completely:*
    (a) $16 - x^2$ (b) $3ax^2 - 3a$
    (c) $2x^2 + 8$ (d) $x^2 + 25$
21. *Find S:*
    (a) $^3\!/\!_7 x = 5$ (b) $^-x = 8$ (c) $4x = 0$
    (d) $x - 8 = {}^-8$ (e) $3x - 3 = 1$
22. *Solve for Q:* $V = Qx + Qy$
23. *Solve for M:* $W = 1 - Mz - My$
24. *Factor:* $8x^2 + 8y^2$ (Be careful.)
25. Give an example of a second-degree equation in one variable.
26. Complete this multiplication by using the distributive axiom:
    $3 \cdot 46 = 3(40 + 6) = ?$
27. Explain the importance of the axioms in a mathematical system such as this course.

Factor completely and check by multiplying:

28. $2x^2 - 12x - 30$
29. $3x^2 - 3$
30. $8x^2 - 16$
31. $8x^2 - 8$
32. $8x^2 + 8$
33. $4xy^2 - 12x$
34. $t^3 - t$
35. $6t^3 - 6t$
36. $12y^2 - 3$
37. $ax + px + qx + bx$

38. $3x^2 - 4$

39. $x^2 + 25$

40. $11x^2 - 121$

41. $36x^2 - 1$

42. $54x^2 - 24$

6-6. THE PRODUCT OF TWO BINOMIALS

The product of two binomials *containing similar terms* is the next special product. An example is $(x + 4)(x + 3)$. We will encounter this type of multiplication so often that it is necessary to memorize the steps leading to the trinomial product.

Consider the following:

$(x + 4)(x + 3)$ can be multiplied by the method shown in section 3-7.

$$
\begin{array}{r}
x + 4 \\
x + 3 \\
\hline
3x + 12 \\
x^2 + 4x \\
\hline
x^2 + (3x + 4x) + 12 = x^2 + 7x + 12
\end{array}
$$

The four terms on the left side of the above equation can be obtained from the following diagram.

1 The product of the first terms, $x \cdot x = x^2$

2 The product of the two inner terms, $4 \cdot x = 4x$

3 The product of the two outer terms, $3 \cdot x = 3x$

4 The product of the last terms, $4 \cdot 3 = 12$

The final product is the sum of these four terms. Notice how steps 2 and 3 give similar terms and their sum is the middle term in the trinomial product, $x^2 + 7x + 12$.

examples:

(a) $(x + 6)(x + 2) = x^2 + 6x + 2x + 12 = x^2 + 8x + 12$

(b) $(x + 6)(x - 2) = x^2 + 6x - 2x - 12 = x^2 + 4x - 12$

(c) $(x - 6)(x - 2) = x^2 - 6x - 2x + 12 = x^2 - 8x + 12$

(d) $(3y - 4)(2y + 3) = 6y^2 - 8y + 9y - 12 = 6y^2 + y - 12$

(e) $(4k - 5)(6k - 5) = 24k^2 - 30k - 20k + 25 = 24k^2 - 50k + 25$

(f) $(k + 6)^2 = (k + 6)(k + 6) = k^2 + 6k + 6k + 36 = k^2 + 12k + 36$

Exercise 32

Multiply as indicated:

1. $(x + 2)(x + 1)$
2. $(x - 3)(x - 2)$
3. $(x + 3)(x + 4)$
4. $(x - 6)(x - 1)$
5. $(x - 8)(x + 2)$
6. $(x - 7)(x + 4)$
7. $(3x - 1)(3x - 2)$
8. $(3x + 4)(3x - 4)$

9. $(x - 5)(2x + 7)$
10. $(x + 3)(x + 3)$
11. $3x(x + 4)$
12. $(7x - 1)(2x + 4)$
13. $3x^2 - 2x(x + 5)$
14. $(5x - 8)(x - 2)$
15. $(6 - 5x)(3 - 4x)$

Factor completely:

16. $9x - 3y$
17. $2x^2 - 4$
18. $4x^2 - 4$
19. $21x^3 - 7x^2 - 7x$
20. $bx^3 - bx$

21. $x^2 + 1$
22. $5x^2 - 25$
23. $ax^2 - ax$
24. $bx^2 - bx$

Multiply:

25. $(3x - 1)(3x - 1)$
26. $(2x + 3)(2x - 3)$
27. $3(3x + 1)(x - 2)$
28. $^-5(x - 1)(x + 2)$
29. $(x + 3)(x + 4)(x + 5)$
30. $(x + 2)(x - 2)(x + 1)$

31. *Find S:* $3x - 8 = 1$
32. *Find S:* $a + 2(a - 4) = 3 - a(5 - {}^-2)$
33. State the commutative axiom for multiplication.
34. Give three irrational numbers.
35. Give three negative integers.
36. If $K = \{a, b, c, d\}$ and $L = \{c, d, e, f\}$, find (a) $K \cup L$ and (b) $K \cap L$.
37. *Multiply:* $(6a + 1)(a - 4)$
38. *Factor:* $9m^2 - 1$
39. *Factor:* $a^2x - x^3$

40. *Multiply:* $(a + b)(a - b)(a + b)(a - b)$

The following problems are squares of a binomial. Perform the multiplication and see if you can determine a rule for this special product:

41. $(x + 3)^2$	44. $(2x + 3)^2$	47. $(5x + 2)^2$
42. $(x - 5)^2$	45. $(3x - 1)^2$	48. $(6x + 1)^2$
43. $(x - 8)^2$	46. $(8x - 1)^2$	49. $(x - 7)^2$

6-7.   THE SQUARE OF A BINOMIAL

To square the binomial $(x + k)$, the method used in section 6-6 can be employed: $(x + k)^2 = (x + k)(x + k) = x^2 + kx + kx + k^2 = x^2 + 2kx + k^2$. We should note that steps 2 and 3 give identical terms, and that these terms are the product of the two terms in the original binomial. The first and last terms are each perfect squares, being the squares of the two terms in the binomial. We now give a brief procedure for squaring the binomial $(x + k)$.

Step 1.   Square the first term: $x \cdot x = x^2$

Step 2.   Find twice the product of the two terms: $2 \cdot k \cdot x = 2kx$

Step 3.   Square the last term: $k \cdot k = k^2$

Step 4.   Find the sum of these three terms

examples: (a) $(x + 5)^2 = x^2 + 10x + 25$

(b) $(x - 5)^2 = x^2 - 10x + 25$

(c) $(2x - 5)^2 = 4x^2 - 20x + 25$

A helpful way to remember that $(x + k)^2$ has a middle term is to consider finding the area of a square whose sides are $(x + k)$ units long. The area of this square is $(x + k)^2$. (See Figure 6-1.)

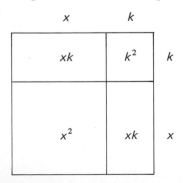

FIGURE 6-1 The square has side $(x + k)$. The area is $(x + k)^2$, which is made up of the two squares and two rectangles in the drawing.

Exercise 33

Square each binomial:

1. $(x + 1)^2$
2. $(x - 2)^2$
3. $(x + 7)^2$
4. $(x + 8)^2$

5. $(2x + 1)^2$
6. $(3x - 2)^2$
7. $(50 + 1)^2$
8. $(9a - 5)^2$

Multiply as indicated:

9. $(x - 3)(x + 3)$
10. $(x + 1)(x - 1)$
11. $(t + 3)(t + 7)$
12. $(t - 4)(t - 3)$
13. $(a + b)^2$

14. $(a - b)^2$
15. $(3x - 5)^2$
16. $(x + 2)(x - 2)(x^2 + 4)$
17. $(x - 6)(x + 6)(x^2 - 36)$
18. $(x + 3)^2(x + 1)$

Factor completely:

19. $am + bm + cm$
20. $ax^2 - ax - a$
21. $3x^2 - 27$

22. $a^2 x^2 - 1$
23. $abx^2 - abx$
24. $abx^3 - abx$

25. What is a solution set?
26. Name the three necessary components of any mathematical system. (See section 1-1.)
27. How many dimes are there in a coin purse where the nickels outnumber the dimes by four and the value of the nickels and dimes is $1.40?
28. Which of the following are irrational numbers?
    $3.66, {}^4/_5, {}^{22}/_7, \sqrt{3}, 5 \cdot \sqrt{3}, {}^-(½), {}^-0.633⅓, \pi$.
29. *Square* $(2x + 7)$.
30. *Square* $(1.2x - 2)$.

Complete each square by choosing the appropriate n:

examples:

(a) Choose a number for n such that $x^2 + 6x + n^2$ is a perfect square. To accomplish this, observe: The middle term, $6x$, is twice the product of x and n. Therefore, divide 6 by 2 and obtain n.
$$x^2 + 6x + n^2 = x^2 + 6x + 3^2 \text{ is a perfect square}$$

(b) $x^2 + 12x + n^2 = x^2 + 12x + 6^2 = x^2 + 12x + 36$

(c) $x^2 - 4x + n^2 = x^2 - 4x + 2^2 = x^2 - 4x + 4$

(d) $x^2 + 3x + n^2 = x^2 + 3x + (\frac{3}{2})^2 = x^2 + 3x + \frac{9}{4}$

31. $x^2 + 2x + n^2$ 35. $x^2 - 7x + n^2$

32. $x^2 - 16x + n^2$ 36. $x^2 + 14x + n^2$

33. $x^2 - 8x + n^2$ 37. $m^2 - 20m + n^2$

34. $x^2 + 5x + n^2$ 38. $k^2 + 13k + n^2$

6-8. FACTORING THE QUADRATIC TRINOMIAL

A polynomial such as $x^2 + 3x + 2$ is called a quadratic or second-degree trinomial. This type of trinomial was encountered in the last two sections as a result of multiplying two binomials containing similar terms.

To factor $x^2 + 3x + 2$, our thinking should proceed as follows:

$$x^2 + 3x + 2$$

$$(x \quad)(x \quad)$$

The first term in the trinomial comes from the two first terms in the binomials.

The second plus sign in the trinomial indicates that the middle signs in the binomials will be alike, that is,

$$(x + ?)(x + ?) \ \text{ or } \ (x - ?)(x - ?)$$

The first plus sign in this trinomial tells us that the two signs in the binomials are + since the middle term is the sum of two terms with like signs.

$$(x + ?)(x + ?)$$

Finally, the factors of 2 (the last term in the trinomial) fill in the last places in the binomials, so that

$$x^2 + 3x + 2 = (x + 2)(x + 1)$$

The above example is the simplest of all quadratic trinomials to factor, since there is but one choice for factors of 2.

To factor $x^2 - x - 6$, the negative sign before the 6 indicates opposite middle signs in the binomials:

$$x^2 - x - 6 = (x + ?) (x - ?)$$

This time we have a choice of 6 and 1 or 3 and 2 to put in place of the question marks.

The four possible pairs of factors are:

$$(x + 1)(x - 6)$$
$$(x + 6)(x - 1)$$
$$(x + 3)(x - 2)$$
$$(x + 2)(x - 3)$$

Each pair of factors gives the correct first and last terms. It is the middle term of the trinomial that dictates which combination is correct. By rapidly checking the middle term obtained from each pair of factors, we find that the last two give a middle term of x but only that last one gives ^-x as a middle term.

Therefore, $x^2 - x - 6 = (x + 2)(x - 3)$

example: Factor: $15x^2 + 14x - 8$

This time the first term $15x^2$ has two possible pairs of factors, $15x$ and x or $5x$ and $3x$. Furthermore, the negative sign in front of the 8 indicates that the middle signs of the binomial factors are opposite. With opposite signs, the middle term is a difference of the inner and outer products. Finally, the $^-8$ can come from $^-(4)(2)$ or $^-(8)(1)$.

Consider all possible pairs of factors:

$(5x + 2)(3x - 4)$	$(5x - 2)(3x + 4)$
$(5x + 4)(3x - 2)$	$(5x - 4)(3x + 2)$
$(5x + 8)(3x - 1)$	$(5x - 8)(3x + 1)$
$(5x + 1)(3x - 8)$	$(5x - 1)(3x + 8)$
$(15x + 2)(x - 4)$	$(15x - 2)(x + 4)$
$(15x + 4)(x - 2)$	$(15x - 4)(x + 2)$
$(15x + 8)(x - 1)$	$(15x - 8)(x + 1)$
$(15x + 1)(x - 8)$	$(15x - 1)(x + 8)$

The first combination gives a middle term of ^-14x, while the first combination in the second column gives the correct pair of factors. This type of factoring requires intelligent trial and error.

Exercise 34

Factor completely and check by multiplication unless otherwise stated:

1. $x^2 + 4x + 3$
2. $a^2 + 6a + 5$
3. $x^2 - 4x + 3$
4. $x^2 - 2x - 8$
5. $y^2 - 3y - 10$
6. $x^2 - 9$
7. $2ax^2 - 8a$
8. $a^2 - 3a - 1$

9. $25 - 16k^2$
10. $3mn - 6mn^2 - 12mn^3$
11. $pqr + prx + pqx$
12. $2x^2 - x - 1$
13. $k^2 - 11k + 28$
14. $12m^2 + 7m - 12$
15. $6x^2 - 25x + 4$

16. *Find S:* $2m + 4m - 3 + 6 = 2(m + 2) + 7$
17. State the associative axiom for multiplication.
18. What is the multiplicative identity?
19. Name three undefined terms used in this book.

20. $2mc^3 - 2mc$
21. $8x^2 - 18x - 18$

22. $6b^2 - 23b + 20$
23. $a^2 - 4a - 5$

24. *Multiply:*
    (a) $2x(3x - 4)$
    (b) $5(x + 3)(x - 3)$
    (c) $(2a + 7)(a - 9)$
    (d) $(2m + 5)^2$

25. $x^2 - 14x + 49$
26. $8a^2 + 24a + 18$
27. $8 - 7a - a^2$

28. $5 + 6a + a^2$
29. $b^2 + 5b - 24$
30. $18x^2 + 18$

31. What is the definition of an axiom?
32. Give two sets whose intersection is empty.

33. $2x^2 - 5x - 12$
34. $9x^2 + 30x + 25$
35. $a^2 - 6a - 18$
36. $x^2 + 7x + 6$

37. If $3x - 8 = {}^-x + 4$, then $S = ?$

38. $30k^2 + 4km - 2m^2$
39. $16a^4 - 25^2$
40. $5x^2 + 13xy + 6y^2$
41. $144y^4 - 9z^2$
42. $3x^2y^2 - 7xyz - 6z^2$
43. $30a^2 + 33ab + 9b^2$
44. $25a^2 - 70ab + 49b^2$

45. $16x^2y^2z^2 - z^2$
46. $3mk^2 - 6km - 144m$
47. $3t - 28 + t^2$
48. ${}^-48x + 9x^2 + 64$
49. $9m^2 - 163m + 18$
50. $2a^4b - 11a^2b + 12b$

Exercise 35*

Evaluate each expression as it is given; then factor the expression and evaluate the factors. The results in each case should be the same.

example: If $x = 4$, evaluate $x^2 - 2x - 3$

$$x^2 - 2x - 3 = (x - 3)(x + 1)$$
$$4^2 - 2(4) - 3 = (4 - 3)(4 + 1)$$
$$16 - 8 - 3 = 1(5)$$
$$5 = 5$$

In the following, $x = 4$ and $y = 3$:

1. $x^2 - 9$
2. $x^2 y - 5xy - 6y$
3. $xy^2 - xy - x$
4. $y^2 + 3y - 18$
5. $x^2 - 8x + 12$

6. $3y^2 - 11y + 8$
7. $5x^2 y^2 + 13xy^2 - 6y^2$
8. $x^2 y^4 + 16x^2$
9. $12y^2 - 2y - 24$
10. $84x^2 - 217x + 140$

6-9.   THE SQUARE OF A TRINOMIAL[†]

To square the trinomial $(a + b + c)$, we may use the long method used in section 3-7.

$$
\begin{array}{r}
a + b + c \\
a + b + c \\
\hline
ac + bc + c^2 \\
ab \quad + bc \quad + b^2 \\
a^2 + ab + ac \\
\hline
a^2 + 2ab + 2ac + 2bc + c^2 + b^2
\end{array}
$$

or $a^2 + b^2 + c^2 + 2ab + 2ac + 2bc$

Examining this product, we see that the first three terms are the squares of the three terms in the trinomial. Each of the remaining three terms is twice the product of any possible pair of terms.

To square a trinomial:

1.   Square each of the three terms.

2.   Take twice the product of all possible pairs of terms. (There are three such pairs.)

3.   Add these six terms.

---

*Optional exercise.
†Optional section.

examples:

(a) $(2a + 3b + 5c)^2 = 4a^2 + 9b^2 + 25c^2 + 12ab + 20ac + 30bc$

(b) $(x - 4y - 3)^2 = x^2 + 16y^2 + 9 - 8xy - 6x + 24y$

Exercise 36

Perform the indicated multiplication:

1. $(a + b + 1)^2$
2. $(c + d + 5)^2$
3. $(a - b - 3)^2$
4. $(2x - 4y + 1)^2$
5. $(5m + p + 4q)^2$
6. $(3k - 2m + 7)^2$
7. $(5p^2 - 2p - 3)^2$
8. $(2m^2 + 3m + 4)^2$
9. Show that $(x - 2y)^2 + (3c)^2$ does not equal $(x - 2y + 3c)^2$
10. $(7x - y + 3)^2 + (3x + y + 7)^2$

6-10. SUMMARY OF THE VOCABULARY OF FACTORING

One of the surest ways to retain the ability to factor is to be able to state in words the types of special products along with their factors. We now summarize all the special products of this chapter.

Factors	*Products*
$a(b + c)$ Use the distributive axiom	$ab + ac$ A polynomial with a common factor
$(x + y)(x - y)$ Multiplying the sum and difference of two terms	$x^2 - y^2$ The difference of two squares
$(x + y)^2$ The square of a binomial	$x^2 + 2xy + y^2$ A trinomial square
$(x + 3)(x + 5)$ Multiplying two binomials that contain similar terms	$x^2 + 8x + 15$ A quadratic trinomial
$(a + b + c)^2$ The square of a trinomial	$a^2 + b^2 + c^2 + 2ab + 2ac + 2bc$ A six-term polynomial

Exercise 37

*Classify each polynomial according to one of the descriptions in
section 6-10, or state that it does not fit any of those categories.
Indicate if there is a common factor:*

examples:

(a) $3x^2 - 3y^2$
   Description: A difference of two squares with
                 a common factor.

(b) $x^2 - 4x + 4$
   Description: Trinomial square.

(c) $(x + y)(2x + 3)$
   Description: None of above categories, as the
                 binomials do not contain similar terms.

1. $3m + 6$
2. $9x^2 - 1$
3. $5x^2 + 5$
4. $8x^2 - 8$
5. $25x^2 - 10x + 1$
6. $x^2 - 8x - 9$
7. $3xy + 2xy^2 - 9xy^3$
8. $(x + 7y)^2$
9. $x^2 + y^2$
10. $3(a^2 - 2)$
11. $2x^2 + 4x + 2$
12. $3x^2 - 9x + 6$
13. $(2x - 3)(4x + 7)$
14. $3(x - 6)(x + 6)$
15. $(3x - 6y + 31)^2$
16. $4x^2 - 9y$

Factor completely:

17. $12x^2 - 75$
18. $a^3 + 5a^2 - 6a$
19. $9y^2 + 36y + 36$
20. $x^4 - 81$

Find S:

21. $9x - 2 = 0$
22. $9x - 2 = {}^-2$
23. $\frac{1}{5}x = 8$
24. $3(x + 2) = 8(x - 4) - 2$

Factor completely:

25. $x^2 + 25$
26. $a^2 - 9a - 24$

6-11. SOLVING EQUATIONS BY FACTORING

We know from experience in arithmetic that any product of a real number and zero is equal to zero. Conversely, if a product is equal to zero, then at least one of its factors must be zero.

THEOREM 6-2: *If $ab = 0$, then $a = 0$ or $b = 0$*

In spite of the obvious nature of this theorem, it is very useful for solving equations in which the polynomial can be factored into linear factors.

A quadratic (second-degree) equation in one variable, like $x^2 + 3x + 2 = 0$, can be written $(x + 2)(x + 1) = 0$. Since the product is zero, it fits into the above theorem as follows:

$$x^2 + 3x + 2 = 0$$
$$(x + 2)(x + 1) = 0$$
$$x + 2 = 0 \text{ or } x + 1 = 0$$
$$x = {}^-2 \text{ or } x = {}^-1$$

Therefore, $S = \{{}^-2, {}^-1\}$

A theorem from advanced algebra tells us that the number of roots of an equation in one variable is equal to the degree of the equation. Again we must rely upon factoring to find these roots.

examples: (a) Find S: $x^2 - 2x = 0$
$$x(x - 2) = 0$$
$$x = 0 \text{ or } x - 2 = 0$$
$$x = 0 \text{ or } x = 2$$
$$S = \{0, 2\}$$

(b) Find S: $2x^3 - 4x^2 - 30x = 0$
$$2x(x^2 - 2x - 15) = 0$$
$$2x(x - 5)(x + 3) = 0$$

According to the theorem of this section, any of the three factors can be zero, that is,

$$2x = 0 \text{ or } x - 5 = 0 \text{ or } x + 3 = 0$$
$$x = 0 \text{ or } \qquad x = 5 \text{ or } \qquad x = {}^-3$$

These are the three roots anticipated from this third-degree equation in one variable, and

$$S = \{0, 5, {}^-3\}$$

Exercise 38

Find S. Factor each polynomial first, if necessary:

1. $x^2 - 4 = 0$
2. $x^2 - 25 = 0$
3. $x^2 - 4x = 0$
4. $x^2 - 6x = 0$
5. $x^2 + 8x - 9 = 0$

6. $(2x - 3)(x + 4) = 0$
7. $(5x + 2)(2x - 1) = 0$
8. $(x - 2)(2x + 7)(3x - 8) = 0$
9. $x(x - 4)(x - 3) = 0$
10. $x^3 - 16x = 0$

Find S:

11. $3x - 4 = 0$
12. $3x^2 - 4x = 0$
13. $2x^2 = 4x$
14. $2x = 4$
15. $5(m - 3) = 3m$

16. $\frac{3}{5}x - 3 = \frac{1}{5}x + 7$
17. $8x^2 - 200 = 0$
18. $4x^2 + 13x + 3 = 0$
19. $2y^2 - 11y + 15 = 0$
20. $8m^2 - 2m - 3 = 0$

21. Show that $3x^2 - 27 = 0$ and $7x^2 - 63 = 0$ have the same solution set, S.
22. Check each of the three roots in example (b), section 6-12, by substitution into the original equation.
23. Why is the following statement false?
    $x^2 - y^2$ is a perfect square.
24. What is the negative (additive inverse) of:

    (a) ⁻6   (b) ⁻½   (c) $\sqrt{2}$ (d) $\dfrac{\sqrt{3}}{3}$

25. What real number is called the multiplicative identity?
26. What are the prime factors of $x^2 + 25$?

Multiply:

27. $(3x + 5)(3x - 5)$
28. $(3x + 5)^2$
29. $(3x + 5)(2x - 5)$

30. $x(x + 3)(x + 4)$
31. $2xy(3x^2 + 2xy + y^2)$
32. $(2y - 7)^2$

33. In problems 27 through 32, which ones are perfect squares?

Find S:

34. $(2x - 1)(x + 7) = 0$
35. $(3x - 2)^2 = 0$
36. $x^2 - x = 2$

Cumulative Review: Chapters 1 through 6

1. If $x = {}^-3$ and $y = {}^-2$, evaluate $3xy - 2y^2$
2. *Subtract* $9x - 3y$ from ${}^-2x + y$.
3. *Find S:* $2 - 3(x + 4) = 2x + 15$
4. *Simplify:* $\dfrac{3^2 - 2 \cdot 4}{6^2 + 5^2}$
5. *Factor completely:*
   (a) $x^2 - 4y^2$ (b) $x^2 - 3x - 4$
   (c) $5x^2 - 50$ (d) $2x^2 + 4x + 2$
6. *Find:* $\{3, 4, 7, 9, 15\} \cap \{2, 4, 6, 11, 15\}$
7. *Factor completely:* $2x^2 + x - 6$
8. *Factor completely:* $3ax^2 - 15ax + 18a$
9. Find three consecutive integers whose sum is 126.
10. Give the number of terms and factors in each entire expression:
    (a) $2xy(3 + x)$ (b) $(x + y)(y + z)$
    (c) $2a + 3x(x + y)$
11. *List:* (a) five rational numbers
    (b) three negative irrational numbers
    (c) four negative integers larger than ${}^-3$
12. *Give an example of*
    (a) a third-degree polynomial in one variable
    (b) a linear equation in one variable
    (c) a linear trinomial in three variables
13. *Multiply:* (a) $(x + y)(x - y)$
    (b) $(3x + y)(2x - 3y)$
    (c) $(2x + 3)^2$
    (d) $3(x - 1)^2$
14. *Simplify:* $3a(a - 4) + a^2 - 2(3 - a^2) + 6a - {}^-2a^2$
15. A man has three times as many nickels as dimes and two more pennies than nickels. The total amount is \$3.10. How many nickels does he have?
16. $(3x - 7)^2 = 9x^2 - 21x + 49$ is false. Why?
17. $(5x + 2)(3x - 8) = 15x^2 + 34x - 16$ is false. Why?
18. Which axiom allows us to remove a common factor from a polynomial?
19. State the commutative axiom for addition.

Find S:

20. $5x - 15 = 0$ 23. $x^2 + 11x + 10 = 0$
21. $5x^2 - 5x = 0$ 24. $25x^2 - 1 = 0$
22. $x(x - 2)(3x + 6) = 0$ 25. $x^2 - 25 = 0$
26. *Solve for m:* $k = mgh - p$
27. *Solve for x:* $A = px + qx$
28. *Solve for y:* $y = mv - y$
29. State three axioms from memory.
30. State three undefined terms used in this book.

7 Rational numbers (fractions)

7-1. VARIOUS FORMS OF THE SAME NUMBER

We have already defined a rational number as a number of the form $\frac{a}{b}$, where a and b are integers and b is not equal to zero. These numbers are conventionally called fractions, and we shall use this term in most cases.

Every fraction has its own algebraic sign, and each numerator and denominator has its algebraic sign. The fraction $\frac{a}{b}$ may be obtained from

$$^+\left(\frac{a}{b}\right) \text{ or } ^-\left(\frac{^-a}{b}\right) \text{ or } ^-\left(\frac{a}{^-b}\right) \text{ or } \frac{^-a}{^-b}$$

As all forms are equivalent, we will choose $\frac{a}{b}$ as standard form and the other three as poor form.

For the fraction $^-\left(\frac{a}{b}\right)$, we may write

$$^-\left(\frac{a}{b}\right) \text{ or } \left(\frac{^-a}{b}\right) \text{ or } \left(\frac{a}{^-b}\right) \text{ or } ^-\left(\frac{^-a}{^-b}\right)$$

As standard form we choose $\frac{^-a}{b}$.

If you study each fraction in the two groups above, you will see that an *equivalent* form of a fraction may be obtained by changing any two signs of one fraction.

For example,
$$^-\left(\frac{^+a}{^-b}\right) = ^+\left(\frac{^+a}{^+b}\right) = \frac{a}{b}$$

According to the definition, $\frac{a}{b}$ means $a(\frac{1}{b})$. More specifically, $\frac{3}{5} = 3(\frac{1}{5})$, $\frac{x}{4} = x(\frac{1}{4})$, $\frac{x+y}{3} = \frac{1}{3}(x + y)$ and $\frac{a}{a+b} = a(\frac{1}{a+b})$. Either form may be used.

examples: Write in standard form.

(a) $\dfrac{1}{^-(x + 2)} = \dfrac{-1}{x + 2}$

(b) $\dfrac{2}{^-x + 3}$ is in standard form, as the entire denominator is not negative.

(c) $\dfrac{^-2}{^-x + 3}$ is in standard form.

(d) $-\dfrac{x}{y} = \dfrac{^-x}{y}$

Exercise 39

Write each fraction in standard form:

1. $\dfrac{2}{-3}$ 2. $\dfrac{^-3}{^-5}$ 3. $\left(\dfrac{^-2}{x}\right)$

4. $\dfrac{x + y}{^-y}$ 5. $^-\left(\dfrac{^-3}{x + 2}\right)$ 6. $^-\left(\dfrac{^-2x}{^-3y}\right)$

Write each fraction as a product:

example: $\dfrac{2x}{y} = 2x\left(\dfrac{1}{y}\right)$

7. $\dfrac{5}{6}$ 8. $\dfrac{4x}{7}$ (Two ways)   9. $\dfrac{x}{x + 2}$

10. $\dfrac{x + y}{x + 4}$ 11. $\dfrac{^-x}{x - 1}$ 12. $\dfrac{^-3}{^-(x - 8)}$

Write each product as a fraction:

13. $3\left(\dfrac{1}{7}\right)$ 14. $^-x\left(\dfrac{1}{x + 2}\right)$ 15. $(x + 2)\left(\dfrac{1}{x}\right)$

16. $(x^2 - 3)\left(\dfrac{1}{x + 2}\right)$ 17. $axy\left(\dfrac{1}{3m}\right)$ 18. $^-2x\left(\dfrac{1}{^-y}\right)$

7-2. REDUCING FRACTIONS

THEOREM 7-1:   *The product of two fractions:*

$$\frac{a}{b} \cdot \frac{c}{d} = \frac{ac}{bd}$$

In words: The product of two or more fractions is a fraction whose numerator is the product of the numerators and whose denominator is the product of the denominators.

With this theorem we may regard $^{35}/_{40}$ as $\frac{7 \cdot 5}{8 \cdot 5} = \frac{7}{8} \cdot \frac{5}{5}$. However, the factor $^5/_5$ is equal to 1, the identity for multiplication. The fraction $^5/_5$ need not appear, as it has no effect upon the product.

The process of identifying 1 as a factor and then choosing not to write it is called cancellation.

The key word in the definition of cancellation is "factor." If the number 1 appears as a *term*, then it cannot be ignored. With this concept of cancellation we proceed to *reduce fractions* in two steps:

1. Factor the numerator and denominator of the fraction.

2. Cancel the identical factors that appear in the numerator and denominator.

examples:

(a) $\dfrac{15}{21} = \dfrac{\cancel{3} \cdot 5}{\cancel{3} \cdot 7} = \dfrac{5}{7}$

(b) $\dfrac{x^2 - y^2}{x^2 - xy} = \dfrac{(x + y)(\cancel{x - y})}{x(\cancel{x - y})} = \dfrac{x + y}{x}$

Note carefully that the x which appears in the numerator is a term and as such cannot be canceled.

(c) $\dfrac{12x^2 y^3 z}{8xy^2 z} = \dfrac{3 \cdot \cancel{4} \cancel{x} x \cancel{y} \cancel{y} y \cancel{z}}{2 \cdot \cancel{4} \cancel{x} \cancel{y} \cancel{y} \cancel{z}} = \dfrac{3xy}{2}$

This is the same type of problem encountered in section 3-8.

(d) $\dfrac{x^2 + 4x - 12}{x^2 - 6x + 8} = \dfrac{(\cancel{x - 2})(x + 6)}{(x - 4)(\cancel{x - 2})} = \dfrac{x + 6}{x - 4}$

The problem of dividing a polynomial by a monomial was deferred from section 3-8 to the present. It will be done in exactly the same way as reducing a fraction, if possible.

examples:

(a) $(3x^2 - 2x) \div x$ becomes

$$\frac{3x^2 - 2x}{x} = \frac{\cancel{x}(3x - 2)}{\cancel{x}} = 3x - 2$$

(b) $\dfrac{2a^2 - 3a - 1}{a}$

As there is no common factor in the numerator of example (b), no division can be performed.

Exercise 40

Reduce to lowest terms:

1. $\dfrac{24}{30}$ 2. $\dfrac{15}{25}$ 3. $\dfrac{72}{81}$ 4. $\dfrac{13}{52}$

5. $\dfrac{3x}{x}$ 6. $\dfrac{2xy}{2x}$ 7. $\dfrac{8k^2 y}{2k}$ 8. $\dfrac{3m^2 n^2 p}{6mnp}$

9. $\dfrac{28p^3 y^2 z^4}{16py^2 z}$ 10. $\dfrac{x(x + y)}{x^2}$ 11. $\dfrac{3(x + y)}{6(x + y)}$ 12. $\dfrac{x^2 - 4}{x + 2}$

13. $\dfrac{(x + 4)^2}{x + 4}$ 14. $\dfrac{x^2 - 4x}{x}$ 15. $\dfrac{x^2 - 1}{x - 1}$ 16. $\dfrac{x^2 - 3x + 2}{x - 1}$

17. $\dfrac{xy - y}{x^2 - 1}$ 18. $\dfrac{x^2 - 2x - 3}{x^2 - 9}$ 19. $\dfrac{3t^3 - 6t}{4t^4 - 8t^2}$ 20. $\dfrac{k^2 + 6k + 9}{k + 3}$

21. $\dfrac{x^2 + 4}{x^2 - 4}$ 22. $\dfrac{x + 3}{3}$ 23. $\dfrac{4x^2 + 12}{4}$ 24. $\dfrac{x^2 + x - 6}{x - 2}$

25. What special name is given to the number 1?
26. Give three rational numbers that are not integers.
27. Give three rational numbers that are integers.
28. *Find S:* $2 - 3(x + 4) = 2$
29. Give two irrational numbers.
30. Give two irrational numbers that are not real numbers.
31. *Factor:* $x^2 + 25$
32. *Factor:* (a) $x^2 - 24x - 25$ (b) $3m^3 - 27m$
33. *Multiply:* (a) $(5y - 4)^2$ (b) $(5y + 4)(5y - 4)$

34. *Reduce:* (a) $\dfrac{3y - 9}{2y - 6}$ (b) $\dfrac{3ab^2 - a}{ab + a}$

35. *Reduce:* (a) $\dfrac{3xy - 2ky - y}{3x - 2k - 1}$ (b) $\dfrac{a - 1}{a^2 - 1}$

36. *Find S:* (a) $(x + 2)(x + 3) = 0$ (b) $x^2 - 5x = 0$

           (c) $\dfrac{3}{4x} = \dfrac{2}{3}$

37. *Subtract:* $9m^2 - 3m - 2$ from $11m^2 - 3$

38. Find two consecutive even integers such that 3 times the first is 16 more than twice the second.

Reduce:

39. $\dfrac{mk - my}{k^2 - y^2}$ 40. $\dfrac{56y^2 z^2}{36yz^2}$

41. $\dfrac{3x^2 - 2x - 1}{12x^2 + 4x}$ 42. $\dfrac{2y^2 - 2y - 1}{2y^2 - 2y}$

43. $\dfrac{a^4 - 81}{a - 3}$ 44. $\dfrac{2p^2 + 11p - 6}{p^2 + 6p}$

45. State the associative axiom for addition.

46. What is meant by a "solution set"?

Reduce:

47. $\dfrac{a^2 + 7a + 10}{a^2 - a - 6}$ 48. $\dfrac{a^2 - 5a + 6}{a^2 - a - 6}$

49. $\dfrac{2a^2 + 12a + 18}{6a^2 - 54}$ 50. $\dfrac{6y^2 - 45y - 24}{6y^2 + 15y + 6}$

7-3. MULTIPLICATION AND DIVISION OF FRACTIONS

Theorem 7-1 shows us how to multiply fractions. From this theorem we can observe that the product of several fractions is itself a fraction, and as such may possibly be reduced. However, reducing fractions requires factoring as a first step, and it will be to our advantage to factor and reduce before multiplying the numerators and denominators.

To illustrate this procedure, consider $\frac{3}{7} \cdot \frac{28}{33}$. If we immediately multiply we get $\frac{84}{231}$, which now must be reduced by factoring. A more direct solution would be to factor first, cancel, then multiply:

$$\frac{3}{7} \cdot \frac{28}{33} = \frac{3}{7} \cdot \frac{4 \cdot 7}{3 \cdot 11} = \frac{\cancel{3} \cdot 4 \cdot \cancel{7}}{\cancel{7} \cdot \cancel{3} \cdot 11} = \frac{4}{11}$$

examples:

(a) $\dfrac{3x}{y} \cdot \dfrac{2y}{9x} = \dfrac{\cancel{3x} \cdot 2\cancel{y}}{\cancel{y} \cdot \cancel{3} \cdot 3\cancel{x}} = \dfrac{2}{3}$

(b) $\dfrac{q^2 - 1}{3q} \cdot \dfrac{15q^3}{q^2 - q} = \dfrac{(q + 1)(\cancel{q-1}) \cdot \cancel{3} \cdot 5\cancel{q} \cdot \cancel{q} \cdot q}{\cancel{3}\cancel{q} \cdot \cancel{q}(\cancel{q-1})}$

$$= 5q(q + 1) = 5q^2 + 5q$$

(c) $\dfrac{4mp}{mp^2 - 9m} \cdot \dfrac{p^2 + 4p + 3}{16m^2 p^3}$

$$= \dfrac{\cancel{2} \cdot 2\cancel{mp}\,(\cancel{p+3})(p + 1)}{\cancel{m}(\cancel{p+3})(p - 3)\,\cancel{2} \cdot 2 \cdot 2 \cdot 2m \cdot m \cdot \cancel{p} \cdot p \cdot p}$$

$$= \dfrac{p + 1}{(p - 3)\,4m^2 p^2} = \dfrac{p + 1}{4m^2 p^3 - 12m^2 p^2}$$

(d) $\dfrac{2t^2 - t - 3}{5t^2 + 2t - 3} \cdot \dfrac{5t^2 + 12t - 9}{t^2 + 6t + 9}$

$$= \dfrac{(2t - 3)(\cancel{t+1})}{(\cancel{5t-3})(\cancel{t+1})} \cdot \dfrac{(\cancel{5t-3})(\cancel{t+3})}{(t + 3)(\cancel{t+3})} = \dfrac{2t - 3}{t + 3}$$

Division has been defined in terms of multiplication, that is, $a \div b$ means $a(\frac{1}{b})$. To divide fractions, then, all we do is multiply the dividend by the reciprocal (multiplicative inverse) of the divisor.

examples:

(a) $4a \div 2 = \dfrac{4a}{1} \cdot \dfrac{1}{2} = 2a$

(b) $\dfrac{rs + s}{r + s} \div \dfrac{r^2 + 2r + 1}{rs^2}$

$$= \dfrac{rs + s}{r + s} \cdot \dfrac{rs^2}{r^2 + 2r + 1}$$

$$= \dfrac{s(r + 1)}{(r + s)} \cdot \dfrac{rs \cdot s}{(r + 1)(r + 1)} = \dfrac{rs^3}{(r + s)(r + 1)}$$

Exercise 41

Perform the indicated operations:

1. $\dfrac{3a}{2} \cdot \dfrac{5x}{7}$ 2. $\dfrac{3a}{2} \div \dfrac{5x}{7}$ 3. $\dfrac{4m}{5n} \cdot \dfrac{25}{28}$

4. $\dfrac{6x^2 y}{3x} \cdot \dfrac{3x^2 y^2}{2x^3 y^3}$ 5. $2 \div \dfrac{1}{a}$ 6. $\dfrac{1}{2a} \div \dfrac{4}{6a}$

7. $\dfrac{a+1}{a^2} \cdot \dfrac{a^3}{(a+1)^2}$ 8. $\dfrac{y^2 - y}{8a} \div \dfrac{y}{2a^2 - a}$

9. $\dfrac{(y-3)}{(y+2)} \cdot \dfrac{y^2 - 4}{y^2 - 3y}$ 10. $\dfrac{m^2 - m - 2}{m^2 + m - 2} \cdot \dfrac{m+2}{m-2}$

11. $\dfrac{5p(p+1)}{10p^2} \cdot \dfrac{8p}{p^2 - 3p - 4}$ 12. $16ax^2 \cdot \dfrac{3a}{4x^2}$

13. $9m^3 \div 3m$ 14. $\dfrac{3k+15}{2k} \cdot \dfrac{4k}{4k+40}$

15. $\dfrac{z-3}{z+4} \div \dfrac{z+4}{z-3}$

16. *Reduce:* $\dfrac{3ax - 12a}{x^2 - 6x + 8}$

17. *Reduce:* $\dfrac{2y^2 + y - 15}{2y^2 + 3y - 20}$

18. *Reduce:* $\dfrac{2y^2 (y)(^-15)}{2y^2 (3y)(^-20)}$

19. *Find S:* $3y - 2 - 4y = {}^-y + 7 + 6y$
20. *Multiply:*   (a) $(3a - 2)(4a + 7)$ (b) $a^2 (3a^2 - 2a + 1)$
21. Which axiom are you using in problem 20(b)?
22. Which axiom is this; $a + (b + c) = (a + b) + c$
23. Count the terms and factors in
     (a) $3a(a + 2)(a + 3)$ (b) $2a(a + 2) + 3$
     (c) $(a + b)(a - b) - (c + d)(c - d)$

24. *Divide:* $\dfrac{3x^2 - 2x - 1}{x^2 - 1}$ by $\dfrac{6x + 2}{2x^2 + 2x}$

25. $\dfrac{2rs + s^2}{r^2 + 2rs} \div \dfrac{rs + s^2}{r^3 + 2r^2 s}$

26. $\dfrac{12 + m - m^2}{9 - m^2} \cdot \dfrac{3 + 2m - m^2}{8 + 2m - m^2} \cdot \dfrac{m + 2}{m^2 + m}$

27. $\dfrac{3p^2 q}{2p - 2q} \div \dfrac{5pq}{5p - 5q}$

28. $\dfrac{k^2 + 5k + 6}{km + 2m} \cdot \dfrac{3}{k + 3}$

29. $\dfrac{2p^2 - 5p - 3}{3p^2 - 10p - 8} \cdot \dfrac{p^2 + p - 12}{p^2 - 9}$

30. $\dfrac{2x^3}{xy + 3} \div \dfrac{4x}{xy + 3}$

31. *Solve for g:* (a) $k = \dfrac{1}{2} gt^2$ (b) $mg = g - t$

32. *Factor:* (a) $3a^2 - 3a - 18$ (b) $16mn^2 - 16m$
33. Find three consecutive integers whose sum is 42.

34. $\dfrac{ax + ay}{12} \cdot \dfrac{24}{a^2}$

7-4. MULTIPLICATION AND DIVISION IN THE SAME EXPRESSION

According to the order of operations (section 2-8), when no parentheses indicate where to begin, multiplication and division are performed in the order of appearance from left to right. A problem with both operations can be changed to one in which only multiplication is involved. This is done by replacing each divisor with its reciprocal, according to the definition of division.

examples: (a) $\dfrac{3}{5} \cdot \dfrac{12}{21} \div \dfrac{4}{5} = \dfrac{3}{5} \cdot \dfrac{12}{21} \cdot \dfrac{5}{4}$

$= \dfrac{3}{\cancel{5}} \cdot \dfrac{\cancel{3}}{\cancel{3}} \cdot \dfrac{\cancel{4}}{7} \cdot \dfrac{\cancel{5}}{\cancel{4}} = \dfrac{3}{7}$

(b) $\dfrac{4x}{5} \div \dfrac{2x}{3} \cdot \dfrac{1}{x} = \dfrac{4x}{5} \cdot \dfrac{3}{2x} \cdot \dfrac{1}{x}$

$= \dfrac{2 \cdot \cancel{2x} \cdot 3 \cdot 1}{5 \cdot \cancel{2x} \cdot x} = \dfrac{6}{5x}$

Exercise 42

Perform the indicated operations:

1. $\dfrac{2}{3} \cdot \dfrac{4}{5} \div \dfrac{12}{15}$

2. $\dfrac{ax}{b} \div \dfrac{ax^2}{b} \cdot \dfrac{3a}{b}$

3. $\dfrac{p^2 q}{a} \cdot \dfrac{pq^2}{p} \div pqa$

4. $\dfrac{a+b}{c} \div \dfrac{a+b}{c^2} \div \dfrac{c}{a}$

5. $\dfrac{x^2 - x}{3} \div \dfrac{x^2 - 1}{9x + 9} \cdot \dfrac{(x+1)^2}{3x^2}$

6. $\dfrac{m - mn}{18 - 3n} \cdot \dfrac{4}{1 - 4n + 3n^2} \div \dfrac{3m^2}{6 - 19n + 3n^2}$

7. $\dfrac{ab}{c} \div \dfrac{a^3 b^2}{c^3} \div \dfrac{a}{c} \cdot \dfrac{c^2}{ac}$

8. $\dfrac{x^2 z^2}{y^3 - y^2 z} \cdot \dfrac{2x + 2z}{xz} \div \dfrac{4xz^3}{x^2 - z^2}$

9. $\dfrac{8y^3 + 8y^2 + 2y}{3y^2 - 22y - 16} \cdot \dfrac{4y^2 - 33y + 8}{4y^2 + 7y - 2} \div \dfrac{4y^2 - 1}{15y^2 + 10y}$

10. $\dfrac{5b^2 - 5b}{3} \div \left(\dfrac{6b - 60}{b^2 - 9b - 10} \div \dfrac{b^2}{2 - 2b^2} \right)$

11. *Find S:* $2(x - 3) + 4(6 - 3x) = {}^{-}12$

12. *Square:* $(2a - 5)$

13. $\dfrac{5x^2 - 10x - 15}{5xy} \cdot \dfrac{x^2 - xy}{x^2 - y^2}$

14. $\dfrac{2x^2}{3y^3} \cdot \dfrac{y^3 + y^2}{x + x^2} \div \dfrac{y^3 - y}{(x+1)^2}$

15. What is the additive identity?

7-5.   LOWEST COMMON MULTIPLE

The lowest common multiple (LCM) of a set of numbers is the smallest positive number that can be divided without remainder by each number of the set. For example, the LCM for {5, 2} is 10. But choose the set {12, 20, 42} and the LCM is not so obvious; so let us proceed in an orderly fashion *to construct the LCM.*

First, find the prime factors of each member of {12, 20, 42}.

$$12 = 2 \cdot 2 \cdot 3$$
$$20 = 2 \cdot 2 \cdot 5$$
$$42 = 2 \cdot 3 \cdot 7$$

Second, form the product of all different prime factors, repeating each one as many times as it occurs in any one member of the set.

$$2 \cdot 2 \cdot 3 \cdot 5 \cdot 7 = 420$$

Our claim is that 420 is the LCM for {12, 20, 42}. Should we try to find a smaller common multiple, one of the prime factors of 420 would be missing and this would mean that one of the members of the given set would not divide without remainder into this smaller number (smaller than 420).

examples: (a)   Find the LCM for {$3x$, $9x$, $21x$}

$3x$ is prime

$9x = 3 \cdot 3x$

$21x = 7 \cdot 3x$

$\text{LCM} = (3x)(3)(7) = 63x$

(b)   Find the LCM for {$x^2 - 4$, $4x$, $x^2 - 4x + 4$}

$$x^2 - 4 = (x + 2)(x - 2)$$
$$4x = 2 \cdot 2x$$
$$x^2 - 4x + 4 = (x - 2)(x - 2)$$
$$\text{LCM} = 2 \cdot 2x(x + 2)(x - 2)(x - 2) \quad \text{or}$$
$$4x(x + 2)(x - 2)^2$$

(c)   Find the LCM for {32, 25, 80, 30}

$$32 = 2 \cdot 2 \cdot 2 \cdot 2 \cdot 2$$
$$25 = 5 \cdot 5$$
$$80 = 2 \cdot 2 \cdot 2 \cdot 2 \cdot 5$$
$$30 = 2 \cdot 3 \cdot 5$$
$$\text{LCM} = 2 \cdot 2 \cdot 2 \cdot 2 \cdot 2 \cdot 5 \cdot 5 \cdot 3$$
$$= 2400$$

Exercise 43

Find the LCM for each set, unless otherwise stated:

1. $\{3, 15\}$ 2. $\{8, 4, 20\}$
3. $\{25, 30, 45\}$ 4. $\{18, 40, 21, 49\}$
5. $\{2x, 8x\}$ 6. $\{3x^2, 12x\}$
7. $\{2, 2a, 2ab\}$ 8. $\{16x^2, 12x^3y, xy^2\}$
9. $\{a - 1, a - 2\}$ 10. $\{ax, a^2x^2 - ax\}$
11. $\{12a, 4a^2 - 4\}$ 12. $\{a^2 - 9, a^2 - 4a + 3\}$

13. What is the effect on an algebraic expression of multiplying it by the multiplicative identity?

14. *Reduce:*   (a) $\dfrac{3x - 9}{4x - 12}$ (b) $\dfrac{a^2 - b^2}{a^2 b - ab^2}$

15. $\dfrac{3x + 3}{4x + 4} \cdot \dfrac{8x - 8}{6x + 6} = ?$

16. (a) $\dfrac{3xy}{4m} \div \dfrac{3x}{m} = ?$ (b) $\dfrac{3x^2 - 75}{3x^2 + 9x - 30} = ?$

17. $\{a, bc, d^2\}$ 18. $\{y^2 - 2y + 1, y^2 - 1\}$
19. $\{y^2, y^2 + 3y\}$ 20. $\{5x - 10, 2x^2 - x - 6\}$
21. $\{m^2 + 5m + 6, m^2 - 9, m^2 + 3m + 2\}$
22. Which *axiom* allows this operation? $\dfrac{k}{k} \cdot m = m$
23. *Simplify:* $x(4x - 3) - 4(x^2 + 2x) - {}^-(x^2 + 3x)$
24. *Find S:* ${}^-7(a + 1) + 2(1 - a) = 13$
25. $\{p^2 - p - 6, p^2 + 7p + 10, p^2 - 6p + 9\}$

7-6. BUILDING FRACTIONS

We know that a fraction like $\frac{12ax}{6ay}$ can be written $\frac{6a \cdot 2x}{6a \cdot y} = \frac{6a}{6a} \cdot \frac{2x}{y} = 1 \cdot \frac{2x}{y} = \frac{2x}{y}$. The symmetric axiom for equality (section 4-3) allows us to reverse the steps, that is, build up a fraction with $1 = \frac{6a}{6a}$ as the *building factor*. In fact, any "building factor" may be used.

For example, $\dfrac{2y}{y} \cdot 1 = \dfrac{2y}{y} \cdot \dfrac{3y}{3y} = \dfrac{6xy}{3y^2}$

If a fraction is to be changed to an equivalent fraction with a different denominator, then the correct building factor must be chosen. A simple way to do this is to factor each denominator to see if the new one contains any factor that is not present in the original denominator. If there is a missing factor, this becomes the numerator and denominator of the building factor.

examples:

(a) Change $\dfrac{3}{5}$ to $\dfrac{?}{20}$

1. Factor each denominator

    5 is prime and $20 = 4 \cdot 5$

2. Note that 4 is the factor missing in the original denominator.

3. Form building factor $\dfrac{4}{4}$ so that

$$\frac{3}{5} \cdot 1 = \frac{3}{5} \cdot \frac{4}{4} = \frac{12}{20}$$

(b) $\dfrac{5x}{x + 1} = \dfrac{?}{x^2 - 1} = \dfrac{?}{(x + 1)(x - 1)}$

Observe that $(x - 1)$ is the missing factor, so that

$\dfrac{x - 1}{x - 1}$ is the building factor.

$$\frac{5x}{x + 1} = \frac{5x}{x + 1} \cdot \frac{(x - 1)}{(x - 1)} = \frac{5x(x - 1)}{(x + 1)(x - 1)}$$

$$= \frac{5x^2 - 5x}{x^2 - 1}$$

(c) $\dfrac{x + 5}{x + 2} = \dfrac{?}{3x^2 + 5x - 2} = \dfrac{?}{(3x - 1)(x + 2)}$

$$\frac{x + 5}{x + 2} \cdot \frac{3x - 1}{3x - 1} = \frac{(x + 5)(3x - 1)}{(x + 2)(3x - 1)} = \frac{3x^2 + 14x - 5}{x^2 + 5x - 2}$$

Exercise 44

Change each fraction to an equivalent fraction with the denominator shown:

1. $\dfrac{5}{6} = \dfrac{?}{24}$
               2. $\dfrac{3x}{8} = \dfrac{?}{40}$

3. $\dfrac{2y}{3x} = \dfrac{?}{27x^2}$

4. $\dfrac{2a}{x-4} = \dfrac{?}{6x-24}$

5. $\dfrac{x+1}{x-2} = \dfrac{?}{x^2-4}$

6. $\dfrac{2y-3}{y+3} = \dfrac{?}{y^2+3y}$

7. $\dfrac{3x^2 y}{5pq} = \dfrac{?}{25p^2 q}$

8. $\dfrac{x+3}{2x} = \dfrac{?}{2x^2+6x}$

9. $3 = \dfrac{3}{1} = \dfrac{?}{6}$

10. $ab^2 = \dfrac{?}{4ab}$

11. *Reduce:*   (a) $\dfrac{ax-a}{bx-b}$ (b) $\dfrac{3x^2-48}{x^2-3x-4}$

12. *Find S:* $2m - 3(m-4) + 16 - 3m = 0$

13. *Multiply:*   (a) $(3x-1)^2$ (b) $(5x+2)(5x-2)$
   (c) $(2a-1)(3x+7)$ (d) $3a(2a-5)^2$

14. *Find S by factoring:*   (a) $x^2 - 4x = 0$
   (b) $2x^2 - x - 1 = 0$ (c) $y^2 + 5y - 50 = 0$

15. (a) $\dfrac{x+2}{x^2-4} = \dfrac{?}{x^3-4x}$ (b) $\dfrac{2a}{a^2+a} = \dfrac{?}{a^3-a}$

16. (a) $\dfrac{a+b}{ax+b} = \dfrac{?}{a^2 x^2 - b^2}$ (b) $\dfrac{ab}{ab^2} = \dfrac{?}{a^2 b^2 + ab^3}$

17. State the distributive axiom.

18. (a) $\dfrac{x^2+y^2}{x+y} = \dfrac{?}{2x^2-2y^2}$ (b) $\dfrac{2}{3} = \dfrac{?}{9y^2-18y+9}$

19. (a) $(^-1)^{23} = ?$ (b) $^-3(^-3)^3 = ?$

20. *Divide (use long division):* $(8x^2 + 6x + 10) \div (x+5)$

21. *Factor:*   (a) $am^3 - am$ (b) $pq^2 - 4pq + 4p$

22. (a) $\dfrac{^-16m^2}{21p} = \dfrac{?}{42p^3}$ (b) $\dfrac{^-(x+y)}{3x-y} = \dfrac{?}{9x^2-12xy+3y^2}$

7-7.   ADDITION OF FRACTIONS

In arithmetic we learned that $^2\!/_5 + {}^1\!/_5 = {}^3\!/_5$ while $^2\!/_5 + {}^2\!/_3$ could not be done as the problem is given. The idea that only fractions with common denominators can be added is true for all real numbers.

THEOREM 7-2: $\dfrac{a}{c} + \dfrac{b}{c} = \dfrac{a+b}{c}$

In words: The sum of two fractions with a common denominator is a fraction whose numerator is the sum of the numerators and whose denominator is the common denominator.

The proof of this theorem is based on the distributive axiom:

$$\frac{a}{c} + \frac{b}{c} = \frac{1}{c}(a+b) = \frac{a+b}{c}$$

examples: (a) $\dfrac{3}{2a} + \dfrac{5}{2a} = \dfrac{8}{2a} = \dfrac{4}{a}$

(b) $\dfrac{x+y}{b} + \dfrac{x-y}{b} = \dfrac{2x}{b}$

(c) $\dfrac{x^2}{x-y} + \dfrac{y^2}{x-y} - \dfrac{2xy}{x-y}$

$$= \frac{x^2 + y^2 - 2xy}{x-y} = \frac{x^2 - 2xy + y^2}{x-y}$$

$$= \frac{(x-y)^2}{x-y} = x - y$$

If two or more fractions are to be added and they do not have a common denominator, then *the LCM of all denominators must be found and called the lowest common denominator (LCD).* * When the LCD is found, each fraction must be changed to an equivalent fraction which has the LCD as its denominator. Finally, we can add the fractions and reduce the sum if possible.

examples: (a) Add $\dfrac{3}{4} + \dfrac{2}{5} + \dfrac{1}{20}$

The LCM for $\{4, 5, 20\}$ is 20; this becomes the LCD

$$\frac{3}{4} = \frac{3 \cdot 5}{4 \cdot 5} = \frac{15}{20}, \frac{2}{5} = \frac{4 \cdot 2}{4 \cdot 5} = \frac{8}{20}, \frac{1}{20} = \frac{1}{20}$$

Finally, $\dfrac{15}{20} + \dfrac{8}{20} + \dfrac{1}{20} = \dfrac{24}{20} = \dfrac{6 \cdot 4}{5 \cdot 4} = \dfrac{6}{5}$

---

*It is not absolutely necessary to find the LCM of the denominator, but if a larger common denominator is used, then a sum will result that will have to be reduced. Many times this reduction cannot be performed using elementary methods.

(b)   Add $\dfrac{ax}{b} + \dfrac{x}{c}$

The LCD is bc

$$\dfrac{ax}{b} = \dfrac{axc}{bc}, \; \dfrac{x}{c} = \dfrac{bx}{bc}$$

$$\dfrac{axc}{bc} + \dfrac{bx}{bc} = \dfrac{axc + bx}{bc}$$

(c)   Add $\dfrac{x + 2}{x + 1} + \dfrac{x - 3}{x + 2} - \dfrac{x^2 - 1}{x^2 + 3x + 2}$

To find the LCD:

$x + 1$ is prime, $x + 2$ is prime,

$x^2 + 3x + 2 = (x + 1)(x + 2)$, and

LCD $= (x + 2)(x + 1)$

$$\dfrac{x + 2}{x + 1} = \dfrac{(x + 2)(x + 2)}{(x + 1)(x + 2)} = \dfrac{x^2 + 4x + 4}{(x + 1)(x + 2)}$$

$$\dfrac{x - 3}{x + 2} = \dfrac{(x - 3)(x + 1)}{(x + 2)(x + 1)} = \dfrac{x^2 - 2x - 3}{(x + 2)(x + 1)}$$

$$\dfrac{^-(x^2 - 1)}{x^2 + 3x + 2} = \dfrac{^-(x^2 - 1)}{(x + 2)(x + 1)} \quad \text{This fraction remains unchanged.}$$

Notice the negative sign which is the sign of the numerator.

Finally,

$$\dfrac{x^2 + 4x + 4}{(x + 1)(x + 2)} + \dfrac{x^2 - 2x - 3}{(x + 1)(x + 2)} + \dfrac{^-(x^2 - 1)}{(x + 1)(x + 2)}$$

$$= \dfrac{x^2 + 4x + 4 + x^2 - 2x - 3 - x^2 + 1}{(x + 1)(x + 2)}$$

$$= \dfrac{x^2 + 2x + 2}{(x + 1)(x + 2)} = \dfrac{x^2 + 2x + 2}{x^2 + 3x + 2}$$

Exercise 45

Perform the operations as indicated:

1. Find the LCM for $\{3x^2, 9x^2 - 1, 3x\}$

2. $\dfrac{3}{7} + \dfrac{a}{7} + \dfrac{2}{7}$

3. $\dfrac{5}{a} + \dfrac{6}{a} - \dfrac{4}{a}$

4. $\dfrac{21m}{p} + \dfrac{13m}{p} - \dfrac{6}{p}$

5. $\dfrac{3a + 1}{a + 1} + \dfrac{2a}{a + 1}$

6. $\dfrac{x^2}{x + 1} + \dfrac{1}{x + 1}$

7. $\dfrac{x^2}{x + 3} - \dfrac{9}{x + 3}$

8. $\dfrac{b^2}{b^2 - 1} - \dfrac{2b}{b^2 - 1} + \dfrac{1}{b^2 - 1}$

9. $\dfrac{y(y + 2)}{y^2} + \dfrac{y - 3}{y}$

10. $\dfrac{2p + 1}{p + 1} - \dfrac{3p - 1}{p + 1} + \dfrac{p}{p + 1}$

11. $\dfrac{5(x + 2)}{(x - 3)(x - 2)} + \dfrac{2(x - 3)}{(x - 3)(x - 2)} - \dfrac{6(2x + 5)}{(x - 3)(x - 2)}$

12. *Reduce:*   (a) $\dfrac{2mq - 3m}{8m^2}$ (b) $\dfrac{a^2 x^2 - 1}{ax - 1}$

13. *Multiply:*   (a) $(3p - 1)(2p + 7)$ (b) $(2p + 7)^2$
      (c) $4x(x + 3)(x - 3)$ (d) $5m(m + 3)(m + 4)$

14. $\dfrac{2a}{a + 1} + \dfrac{2}{a + 1}$

15. $\dfrac{3a^3}{9a} + \dfrac{6a^2}{9a} + \dfrac{9a}{9a}$

16. *Find S:* $^-8(x + 4) + 2(x - 3) = {}^-3(x + 5) + 7$

17. *Find the missing numerator:*

   (a) $\dfrac{2y}{x + 2} = \dfrac{N}{x^2 + 2x}$ (b) $\dfrac{x + 2}{x - 5} = \dfrac{N}{x^2 - 25}$

18. $\dfrac{3}{4} + \dfrac{1}{2}$

19. $\dfrac{2x}{3} - \dfrac{4x}{5}$

20. $\dfrac{a}{2} + \dfrac{a}{2x}$

21. $\dfrac{y + 1}{y} - \dfrac{y + 1}{3y}$

22. $\dfrac{pq}{3} + \dfrac{p}{q}$

23. $\dfrac{a}{b} + \dfrac{c}{d}$

24. $\dfrac{ax}{m} + \dfrac{x}{m} + \dfrac{a}{xm}$

25. $p + \dfrac{p}{q}$

26. $\dfrac{3}{x} - 2$

27. $\dfrac{ab}{b} + \dfrac{a^2}{b^2}$ /

28. $\dfrac{y + 2}{y + 3} + \dfrac{y}{2y + 6}$

29. $\dfrac{y - 3}{3} - \dfrac{y + 3}{3y}$

30. $\dfrac{2(a + b)}{3a} + \dfrac{3(a - b)}{2a}$

31. $\dfrac{1}{x + 1} + \dfrac{x + 1}{1}$

32. $a + b + \dfrac{a}{b}$

33. $\dfrac{1}{x} + \dfrac{1}{x^2} + \dfrac{1}{x^3}$

34. $\dfrac{1}{p + q} + \dfrac{1}{p - q}$

35. $\dfrac{1}{p^2 - pq} + \dfrac{1}{p}$

36. $\dfrac{x - y}{xy} - \dfrac{y - z}{yz}$

37. $\dfrac{2y}{x - y} + \dfrac{3y}{x + y}$

38. Demonstrate the commutative axiom for addition with $x(x + y)$.

39. Using $x(x + y)$, demonstrate the commutative axiom for multiplication.

40. What is the intersection of $\{2, 6, 10, 14\}$ and $\{5, 10, 15, 20\}$?

41. Give the additive and multiplicative inverse of ½.

42. *Find S:* (See Exercise 38, section 6-12.)

     (a) $x^2 - 3x = 0$ (b) $x^2 + 4x + 3 = 0$

Add:

43. $\dfrac{3}{3a + 4} - \dfrac{5}{a + 4}$

44. $\dfrac{xy}{xy + 1} + \dfrac{1}{x^2 y^2 - 1}$

45. $\dfrac{p + 3}{p^2 + 5p} + \dfrac{p + 1}{2p + 10}$

46. $\dfrac{x}{x^2 - 16} - \dfrac{3x}{(x - 4)^2}$

47. $\dfrac{x + 2}{x^2 - 3x - 4} + 3$

48. $1 - \dfrac{2}{p + 3} - \dfrac{4}{p - 3}$

49. $\dfrac{2}{2x^2 - x - 3} + \dfrac{5}{6x^2 - x - 12}$

50. $\dfrac{2x + 1}{6x^2 - 11x + 4} - \dfrac{3x - 2}{2x^2 + 9x - 5}$

7-8.   COMPLEX FRACTIONS

A fraction whose numerator or denominator or both contain a fraction is called a complex fraction. For example,

$$\frac{x + \dfrac{1}{x}}{3}, \quad \frac{x + 1}{1 - \dfrac{1}{x}}, \quad \frac{\dfrac{1}{a} + \dfrac{1}{4}}{a + \dfrac{1}{3}}$$

are complex fractions. Fractions that are not complex are called *simple fractions*.

Each complex fraction must be changed into an equivalent simple fraction. To make this change, we shall multiply the complex fraction by 1, in the form of the fraction k/k. In the fraction k/k, k will be the LCD for all of the denominators in the entire complex fraction.

Taking the complex fraction $\dfrac{x + \frac{1}{x}}{3}$, we can regard it as $\dfrac{\frac{x}{1} + \frac{1}{x}}{\frac{3}{1}}$ where the three denominators are 1, x, and 1 and the LCD is x.

Now multiply:

$$\frac{x}{x} \cdot \frac{\left(x + \dfrac{1}{x}\right)}{3} = \frac{x \cdot x + x\left(\dfrac{1}{x}\right)}{x \cdot 3}$$

$$= \frac{x^2 + 1}{3x}, \text{ which is a simple fraction}$$

examples:

(a)   Simplify: $\dfrac{\dfrac{3}{5}}{\dfrac{2}{3}}$

$\dfrac{\dfrac{3}{5}}{\dfrac{2}{3}}$ has an LCD of 15 for all denominators

$$\frac{15}{15} \cdot \frac{\dfrac{3}{5}}{\dfrac{2}{3}} = \frac{15\left(\dfrac{3}{5}\right)}{15\left(\dfrac{2}{3}\right)} = \frac{3 \cdot 3}{5 \cdot 2} = \frac{9}{10}$$

(b) Simplify:
$$\frac{x + y + \dfrac{1}{x}}{\dfrac{1}{2} - \dfrac{1}{x} - \dfrac{1}{2x}}$$

The LCD is $2x$

$$\frac{2x}{2x} \cdot \frac{\left(x + y + \dfrac{1}{x}\right)}{\left(\dfrac{1}{2} - \dfrac{1}{x} - \dfrac{1}{2x}\right)} = \frac{2x\left(x + y + \dfrac{1}{x}\right)}{2x\left(\dfrac{1}{2} - \dfrac{1}{x} - \dfrac{1}{2x}\right)}$$

$$= \frac{2x \cdot x + 2x \cdot y + 2x\left(\dfrac{1}{x}\right)}{2x\left(\dfrac{1}{2}\right) - 2x\left(\dfrac{1}{x}\right) - 2x\left(\dfrac{1}{2x}\right)} = \frac{2x^2 + 2xy + 2}{x - 2 - 1}$$

$$= \frac{2x^2 + 2xy + 2}{x - 3}$$

(c) Simplify:
$$\frac{b + \dfrac{1}{b - 1}}{b - \dfrac{1}{b + 1}}$$

The LCD is $(b - 1)(b + 1)$

$$\frac{(b - 1)(b + 1)}{(b - 1)(b + 1)} \cdot \frac{\left(b + \dfrac{1}{b - 1}\right)}{\left(b - \dfrac{1}{b + 1}\right)}$$

$$= \frac{(b - 1)(b + 1)b + (b - 1)(b + 1)\left(\dfrac{1}{b - 1}\right)}{(b - 1)(b + 1)b - (b - 1)(b + 1)\left(\dfrac{1}{b + 1}\right)}$$

$$= \frac{(b - 1)(b + 1)b + (b + 1)}{(b - 1)(b + 1)b - (b - 1)} = \frac{(b^2 - 1)b + (b + 1)}{(b^2 - 1)b - (b - 1)}$$

$$= \frac{b^3 - b + b + 1}{b^3 - b - b + 1} = \frac{b^3 + 1}{b^3 - 2b + 1}$$

Exercise 46

Simplify each complex fraction (remember to reduce the simple fraction if possible):

1. $\dfrac{\dfrac{3}{7}}{\dfrac{1}{2}}$

2. $\dfrac{\dfrac{2}{3}}{3}$

3. $\dfrac{1}{\dfrac{5}{6}}$

4. $\dfrac{\dfrac{2a}{3}}{\dfrac{3a}{4}}$

5. $\dfrac{\dfrac{a^2}{b}}{\dfrac{b^2}{a}}$

6. $\dfrac{\dfrac{pq}{a^2}}{\dfrac{pq}{3a}}$

7. $\dfrac{1+\dfrac{1}{b}}{2+\dfrac{1}{b}}$

8. $\dfrac{1}{1-\dfrac{1}{x}}$

9. $\dfrac{a+\dfrac{b}{a}}{2a+\dfrac{b}{a}}$

10. $\dfrac{5+\dfrac{1}{x+3}}{3+\dfrac{1}{x+3}}$

11. $\dfrac{\dfrac{1}{a}+\dfrac{1}{b}}{\dfrac{2}{a}+\dfrac{2}{b}}$

12. $\dfrac{ax-\dfrac{1}{x-1}}{\dfrac{1}{x-1}-x}$

Add:

13. $\dfrac{1}{x}, \dfrac{3}{x}$ and $\dfrac{3}{2x}$

14. $\dfrac{5}{x-1} - \dfrac{2}{x} = ?$

15. *Reduce:* (a) $\dfrac{4x^2-1}{6x-3}$ (b) $\dfrac{ab+a}{b^2+2b+1}$

16. *Multiply:* (a) $(x+3)^2$ (b) $(x+3)(x-3)$
     (c) $(x+3)(x+4)$

17. *Find S:* $(3x-2)(2x+5) = 0$

18. *Find S:* $3t = {}^-4t$

Simplify:

19. $\dfrac{3\left(\dfrac{1}{a}\right) + \dfrac{1}{2}}{4\left(\dfrac{1}{a^2}\right) + 3}$

20. $\dfrac{5\left(\dfrac{-3}{4}\right) + 2}{5 - 2\left(\dfrac{-3}{4}\right)}$

21. $\dfrac{\dfrac{1}{x} - \dfrac{1}{x^2}}{1 - \dfrac{1}{x^2}}$

22. $\dfrac{\dfrac{a}{b} + \dfrac{a}{b+1}}{3}$

23. $2 + \dfrac{1 - \dfrac{1}{3}}{1 + \dfrac{1}{3}}$

24. $\dfrac{a + b}{\dfrac{1}{a} + \dfrac{1}{b}} - \dfrac{\dfrac{1}{a} + \dfrac{1}{b}}{a + b}$

7-9. EVALUATION OF EXPRESSIONS CONTAINING FRACTIONS

In preparation for checking solutions to the equations in section 7-10, we present the next exercise. Make certain that each variable is replaced before any arithmetic is done.

examples:

(a) If $x = {}^-1$, $y = {}^-3$, and $z = \dfrac{1}{4}$, evaluate $2xy^2 - \dfrac{3}{4}x^2 z$

$$2xy^2 - \frac{3}{4}x^2 z$$

$$= 2({}^-1)({}^-3)^2 - \frac{3}{4}({}^-1)^2 \left(\frac{1}{4}\right)$$

$$= 2({}^-1)(9) - \frac{3}{4}(1)\left(\frac{1}{4}\right)$$

$$= {}^-18 - \frac{3}{16}$$

$$= {}^-18\frac{3}{16} \quad \text{or} \quad \frac{{}^-291}{16}$$

(b) If $x = 3$ and $y = 6$, evaluate $\dfrac{1}{x - 3} + \dfrac{y}{2}$

$$\frac{1}{x - 3} + \frac{y}{2}$$

$$= \frac{1}{3 - 3} + \frac{6}{2}$$

$$= \frac{1}{0} + 3$$

Since $\frac{1}{0}$ appears in this expression, there is no meaning to the expression. By definition of a rational number, the denominator must never equal zero.

Exercise 47

Evaluate the following expressions if $x = {}^{-}1$, $y = {}^{-}3$, and $z = \dfrac{1}{4}$:

1. $\dfrac{3y}{4}$

2. $\dfrac{3(x - z)}{x}$

3. $\dfrac{2x}{3} - \dfrac{x}{2}$

4. $2y - \dfrac{y}{5} + 2$

5. $\dfrac{1}{2}z + 2z - 3$

6. $\dfrac{1}{x} + \dfrac{1}{y} + 1$

7. $\dfrac{x^2}{3x} - \dfrac{2x}{3} + \dfrac{1}{3}$

8. $\dfrac{1}{x - 1} + \dfrac{1}{x + 2}$

9. $\dfrac{z}{x - 1} + \dfrac{2}{z + 1}$

10. *Simplify:*   (a) $\dfrac{3 + \dfrac{1}{2}}{4 - \dfrac{1}{3}}$ (b) $\dfrac{1 + \dfrac{1}{a}}{a + 1}$

11. *Add:*   (a) $\dfrac{1}{3a} + \dfrac{1}{2a}$ (b) $\dfrac{xy}{2} - \dfrac{x}{3} - \dfrac{1}{x}$

12. *Reduce:*   (a) $\dfrac{m^2 - n^2}{m^2 - mn}$ (b) $\dfrac{p^2 - 2p - 3}{p^2 - 3p - 4}$

13. If $a = \frac{1}{2}$ and $b = {}^{-}3$, evaluate $9a^2 b^3 + 10a^3 b^4$.

14. *Find S:* $x^2 - 5x + 6 = 0$

Evaluate the following expressions if $x = {}^{-}4$, $y = 0$, and $z = \dfrac{1}{2}$:

15. $\dfrac{3x}{2x + 1} - \dfrac{x}{y + z}$

16. $\dfrac{xyz}{x^2 + y^2} - 3$

17. $\dfrac{1}{2z + x} + \dfrac{x^2}{3}$

18. $\dfrac{xz}{6} - \dfrac{y}{2} - \dfrac{3z}{x}$

19. $\dfrac{1}{x^2} + \dfrac{1}{z^2}$

20. $2z^2 - 3z - 4$

7-10. EQUATIONS CONTAINING FRACTIONS

The multiplication axiom for equality allows us to multiply both sides of an equation by any real number. If we choose the LCD of all denominators of the fractions in an equation as the number by which to multiply each side, then we can remove all fractions from the equation. Without fractions the equation is much easier to solve.

examples: (a) Find S: $\dfrac{x}{3} + 2 = \dfrac{1}{2}$

Solution: The LCD is 6. Multiply each side by 6.

$$6\left(\frac{x}{3} + 2\right) = 6\left(\frac{1}{2}\right)$$

$$6\left(\frac{x}{3}\right) + 6(2) = 3$$

$$2x + 12 = 3$$

$$2x = {}^-9$$

$$x = \frac{{}^-9}{2}$$

$$S = \left\{\frac{{}^-9}{2}\right\}$$

(b) Find S: $\dfrac{x}{x + 1} - \dfrac{1}{2} = \dfrac{3}{4}$,

where $x \neq {}^-1$

(to make certain denominator is not zero)

Solution: $4(x + 1)$ is LCD

$$\frac{x}{x + 1} - \frac{1}{2} = \frac{3}{4}$$

$$4(x + 1)\left[\frac{x}{x + 1} - \frac{1}{2}\right] = 4(x + 1)\frac{3}{4}$$

$$4(x + 1)\left(\frac{x}{x + 1}\right) - 4(x + 1)\left(\frac{1}{2}\right) = (x + 1)3$$

$$4x - 2(x + 1) = 3x + 3$$
$$4x - 2x - 2 = 3x + 3$$
$$2x - 2 = 3x + 3$$
$$^-x = 5$$
$$x = ^-5$$
$$S = \{^-5\}$$

Exercise 48

Find S, unless otherwise stated:

1. $\dfrac{x}{3} = 5$

2. $\dfrac{2x}{5} = 3$

3. $x + \dfrac{1}{2} = 3$

4. $2x - 1 = \dfrac{1}{4}$

5. $3x - \dfrac{1}{3} = x$

6. $9x - \dfrac{1}{2} - \dfrac{3}{4} = \dfrac{13}{4}$

7. $\dfrac{x}{4} - 1 = \dfrac{2x}{5} - \dfrac{1}{10}$

8. $\dfrac{x - 3}{4} + \dfrac{x + 3}{4} = 0$

9. $\dfrac{2}{5} x + 6 = 4$

10. $\dfrac{5}{x} = 10, \ x \neq 0$

11. $8 - \dfrac{4}{x} = 7, \ x \neq 0$

12. $\dfrac{2}{3}x + 3 = \dfrac{1}{2}x + 5$

13. $\dfrac{4}{k} - 3 = \dfrac{5}{2k + 3}, \ \dfrac{k \neq 0}{k \neq \dfrac{^-3}{2}}$

14. $\dfrac{p - 5}{8p} = \dfrac{3}{p + 5}, \ \dfrac{p \neq 0}{p \neq ^-5}$

15. *Simplify:*   (a) $\dfrac{pq + \dfrac{1}{8}}{p - 8}$ (b) $\dfrac{x - \dfrac{1}{x}}{x - \dfrac{1}{x^2}}$

16. If two factors have a product equal to zero, what is true of the factors?

17. *Reduce:*   (a) $\dfrac{r - 3}{r^2 - 9}$ (b) $\dfrac{6a + 3b - 3c}{8a + 4b - 4c}$

18. *Add:*   (a) $\dfrac{3}{x} + 2$ (b) $\dfrac{1}{x} + \dfrac{1}{y} + \dfrac{1}{z}$

19. $\dfrac{y + 8}{4} = \dfrac{5y - 2}{6}$

20. $\dfrac{k - 2}{5k} = \dfrac{1}{6} - \dfrac{2}{15k}$, $k \neq 0$

21. $\dfrac{6}{2y + 5} = \dfrac{4}{y - 3}$, $\begin{array}{l} y \neq \dfrac{^-5}{2} \\ y \neq 3 \end{array}$

22. $\dfrac{4}{k + 2} = \dfrac{9}{k - 3}$, $\begin{array}{l} k \neq {}^-2 \\ k \neq 3 \end{array}$

23. $8(x + 3) - 4(x + {}^-2x - 5) - 3(2 - x) = x + 3(x - 8)$

24. $x^2 - 4x - 21 = 0$

25. $\dfrac{2}{x + 2} + \dfrac{3}{(x + 2)(x + 3)} = \dfrac{2}{x + 3}$, $\begin{array}{l} x \neq {}^-2 \\ x \neq {}^-3 \end{array}$

26. *Name each axiom:*   (a) $a(b + c) = ab + ac$
                               (b) $mp = pm$
                               (c) $a + (b + c) = (a + b) + c$

27. Distinguish an axiom from a theorem.

28. List four irrational numbers.

29. $\dfrac{3y - 2}{3} + \dfrac{2y - 1}{2} = \dfrac{4y + 1}{6} - \dfrac{2y - 2}{12}$

30. $\dfrac{4}{x - 5} + \dfrac{3}{x^2 - x - 20} = \dfrac{7}{x + 4}$, $x \neq 5$, $x \neq {}^-4$

31. $\dfrac{2x}{x^2 - 1} + \dfrac{3x}{x^2 - 2x + 1} = \dfrac{5}{x + 1}$, $x \neq 1$, $x \neq {}^-1$

32. $\dfrac{2x + 11}{x + 4} + \dfrac{x - 2}{x - 4} - \dfrac{12}{x^2 - 16} = \dfrac{7}{2}$

33. $\dfrac{3}{4}(x + 1) + (x + 2) = 50$

34. $.05x + .01(x + 2) = 3.62$

35. $\dfrac{2}{3}(x + 3) - \dfrac{1}{2}(x + 2) = 1$

36. $\dfrac{4}{5}(x - 8) + \dfrac{1}{3}(2x + 3) = \dfrac{7}{15}$

7-11. MORE STATED PROBLEMS

In section 4-8, examples of stated problems were given. It would be a good idea to review similar examples at this time. In addition, two new types of problems will be presented.

examples: (a)   At present Tim is 3 times as old as Steve. Three years from now Tim will be twice as old as Steve. How old is each now?

	Age Now	*Age in 3 Years*
Steve	x(years)	$x + 3$
Tim	$3x$	$3x + 3$

(Notice how each age is increased by 3 years regardless of the relationship between their present or future ages.)

Now use the final piece of information:

Tim's age in 3 years = 2(Steve's age in 3 years)

$$3x + 3 = 2(x + 3)$$
$$3x + 3 = 2x + 6$$
$$x = 3$$

Answer: Steve's age now = x = 3 years
Tim's age now = $3x$ = 9 years

Check: In 3 years Tim will be 12 and Steve will be 6, which will make Tim twice as old as Steve.

(b)   Consider the following drawing of a simple balance board (lever) and pivot point (fulcrum). The teeter-totter is an example of this.

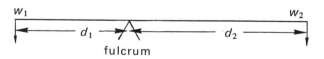

fulcrum

In order to maintain equilibrium (no movement), the product $w_1 d_1$ must equal $w_2 d_2$, where w_1 and w_2 represent weights placed on the balance and d_1, d_2 are the respective distances from the fulcrum.

Let w_1 = 80 pounds, w_2 = 60 pounds, and d_1 = 6 feet. How far from the fulcrum must w_2 be placed in order to balance w_1? The equation

$$w_1 d_1 = w_2 d_2$$

applies here with d_2 being the only unknown. Substituting, we get (80 pounds)(6 feet) = (60 pounds)(d_2 feet). Assuming d_2 to be in feet, we can write

$$80(6) = 60d_2$$

$$\frac{480}{60} = d_2$$

$$d_2 = 8$$

Answer: Place w_2 8 feet from the fulcrum on the opposite side from w_1.

Check: $w_1 d_1$ = (80 pounds)(6 feet) = 480 foot-pounds

$w_2 d_2$ = (60 pounds)(8 feet) = 480 foot-pounds

(c) As an extension of the theory presented in example (b), let us consider the case when two or more weights or forces are applied to one side of the fulcrum.

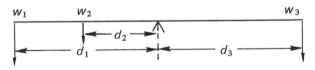

To attain balance, the sum $w_1 d_1 + w_2 d_2$ must equal $w_3 d_3$.

Suppose a 300-pound rock is to be moved with the aid of a board used as a lever pivoting on a smaller piece of wood 2 feet from the large rock. (See drawing.)

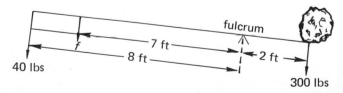

One man is 8 feet from the fulcrum (small piece of wood) and is pushing down on the lever with a force of 40 pounds. A second man is needed to move the rock. If this man can push on the lever at a point one foot closer to the rock, how much force (f) must he apply to start the rock moving?

The formula: $w_1 d_1 + w_2 d_2 = w_3 d_3$ applies.

Let w_1 = 40 pounds, d_1 = 8 feet, $w_2 = f$,

d_2 = 7 feet,

w_3 = 300 pounds, and d_3 = 2 feet

$$40(8) + f(7) = 300(2)$$
$$320 + 7f = 600$$
$$7f = 280$$
$$f = 40$$

Answer: The second man must apply 40 pounds of force.

Exercise 49

For each problem write an equation and solve it. Label all answers. Make a sketch when possible:

1. If ¾ of a number is 21, find the number.
2. If ½ a number is increased by ⅔ of the same number, the sum is equal to 49. Find the number.
3. Anne has 2½ times as many dimes as nickels. If the value of her coins is $5.40, how many dimes does she have?
4. David invested some money at 6% interest. At the end of one year his investment, with interest, was worth $84.80. How much did he originally invest?
5. Find two numbers whose sum is 40 if 1½ times the smaller equals the larger.
6. Find three consecutive even integers such that the first plus ½ the second minus the third equals 13.
7. Smith's Candy Company sold ¾ as many 8¢ candy bars as 5¢ bars one year. If total sales were $3960, how many of each bar were sold?

8. Beth is ⅗ as old as Becky. In 7 years Beth will be ⅔ the age of Becky. How old is each now?

9. *Reduce each fraction:* (a) $\dfrac{mx - x^2}{m - x}$ (b) $\dfrac{ab - b}{a^2 - 1}$

                (c) $\dfrac{a^2 - 3a - 4}{a^2 - a - 12}$ (d) $\dfrac{m^2 x^2 - 4x^2}{amx + 2ax}$

10. *Add:*   (a) $\dfrac{3}{x} + \dfrac{1}{x} + \dfrac{10x}{x^2}$ (b) $2 + \dfrac{1}{a}$

         (c) $\dfrac{3}{a + 1} - \dfrac{2}{a - 1}$ (d) $\dfrac{a + 2}{2a} - \dfrac{a - 3}{6a}$

11. Two boys are balanced on a seesaw. The 90-pound boy is 6 feet from the fulcrum, and the other boy is 9 feet from the fulcrum. How heavy is the other boy?

12. A technician wishes to balance a 120-pound weight and a 100-pound weight. The 100-pound weight is placed 40 feet from the fulcrum. Where must the 120-pound weight be placed?

13. Two men are on opposite ends of a 10-foot teeter-totter with the fulcrum in the middle. The man on the left weighs 200 pounds while the man on the right weighs 160 pounds. To balance the two men, a 50-pound boy is placed on the teeter-totter. Where must he be placed?

14. *Find S:*   (a) $(3x - 6)(x + 2) = 0$ (b) $3x - 8 = 1$
15. *Factor:*   (a) $a^3 b^2 - a^2 b^2 - ab^3$ (b) $16x^2 - 9$
            (c) $3m^2 - 2m - 1$ (d) $ax^2 - ay^2$

16. John has half as much money as Brad. Loren has ⅓ as much money as Brad. Together they have $275. How much does Loren have?

17. A certain number is subtracted from the numerator of ⁴⁄₇. This same number is added to the denominator, and the resulting fraction is equal to 10. Find that number.

18. In a recent attitude survey, ½ of those questioned said "no," and ⅖ said "yes." The remaining 30 students had no opinion. How many were surveyed?

19. A dry mixture for concrete calls for 3 parts gravel, 2 parts sand, and one part cement. If a batch weighs 6600 pounds, what is the weight of the sand?

20. At present, the ages of Dotty and Susie total 60 years. Twelve years ago Dotty's age was ⅘ of Susie's age. How old is each now?

7-12.   RATIO AND PROPORTION*

DEFINITION: *A* ratio *is a fraction used to compare two or more quantities.*

If a mixture of salt and pepper calls for 5 parts salt to 2 parts pepper, then the *ratio of salt to pepper is 5 to 2* and is written ⅝ or 5:2. The fraction expressing the ratio should be reduced to lowest terms, and the numerator and denominator should have the same units. When comparing 8 feet to 2 yards, the ratio 4 to 3 is obtained by changing 2 yards to 6 feet and then reducing the fraction ⅚ to ⁴⁄₃.

example: What is the ratio of 2 pounds to 2 ounces?

Solution: 2 pounds = 32 ounces

2 pounds to 2 ounces becomes

$$\frac{32 \text{ ounces}}{2 \text{ ounces}} = \frac{16}{1} \text{ or 16 to 1}$$

Note: There is *no label* on the ratio itself.

DEFINITION: *A* proportion *is a statement that two ratios are equal.*

The equation *a/b = c/d* is a proportion and is read: *a* is to *b* as *c* is to *d*. In this proportion, *a*, *b*, *c*, and *d* are called the *terms of the proportion*, or each is *a proportional.* More specifically, *a* is the first proportional, *b* is the second proportional, and so forth. When the second and third proportionals are equal, each one is called the *mean proportional.*

In the proportion *a/x = x/b*, *x* is the mean proportional between *a* and *b*. When a proportion has a mean proportional, the last term is called the *third proportional.*

When treating the proportion *a/b = c/d* as an equation, we may clear fractions and obtain *ad = bc.* The product *ad* contains the first and last terms, which are called the *extremes* of the proportion, while the product *bc* contains the two middle terms, called the *means* of the proportion.

THEOREM 7-3:   *In a proportion, the product of the extremes is equal to the product of the means.*

$$\frac{a}{b} = \frac{c}{d} \text{ is equivalent to } ad = bc$$

---

*Optional section.

examples: (a) Find the fourth proportional to 1, 2, and 6.

Let x = the fourth proportional

Then $\dfrac{1}{2} = \dfrac{6}{x}$ and $1 \cdot x = 2 \cdot 6$ by Theorem 7-3.

Continuing: $1 \cdot x = 2 \cdot 6$

$x = 12$, the fourth proportiona

(b) Find the third proportional to 2 and 8.

Let x = the third proportional

(8 is the mean proportional)

$\dfrac{2}{8} = \dfrac{8}{x}$ or $2x = 64$

$x = 32$, the third proportional

Exercise 50

1. *What is the ratio of:*
   - (a) 3 pounds to 2 pounds
   - (b) 8 feet to 4 feet
   - (c) 9 inches to 1 foot
   - (d) 11 weeks to 3 days
   - (e) ½ inch to ¾ inch
   - (f) 4⅓ yards to 6½ yards
   - (g) 6 quarters to 3 dimes
   - (h) $5 to 50¢
   - (i) one square foot to one square yard
   - (j) 5 gallons to 6 quarts

2. *Write as a proportion:*
   - (a) 3 is to 7 as 9 is to 21
   - (b) 5 is to 3 as 7 is to x
   - (c) $(x + y)$ is to $x^2 + y^2$ as $3x + 3y$ is to k
   - (d) $5:8 = 6:m$

3. *Using Theorem 7-3, find S:*

   (a) $\dfrac{3}{x} = \dfrac{12}{28}$ (b) $\dfrac{x}{8} = \dfrac{30}{40}$

   (c) $4:7 = x:2$ (d) $a:4 = 5:24$

    (e) $5:x = 30:18$ (f) $9:x = x:4$

    (g) $2:x = x:32$ (h) $y:3 = 8:15$

    (i) $(x + 1):3 = 2:9$ (j) $4:x - 3 = 16:24$

    (k) $2:3 = 3:(x + 1)$ (l) $(x + 1):(x^2 - 1) = 1:2$

4. Find the third proportional to 1 and 3.

5. Find the fourth proportional to 7, 9, and 49.

6. Find the mean proportional between 9 and 16.

7-13. PROBLEM SOLVING WITH PROPORTION*

    Certain stated problems employ the concepts of ratio and proportion.

examples: (a)   A mixture of concrete calls for gravel, sand, and cement in a ratio of 3:2:1. If 12 tons is to be mixed, how much of each material should be used?

        Let x = one unit of the mixture

        Then $3x$ is amount of gravel

                 $2x$ is amount of sand

                  x is amount of cement

$$3x + 2x + x = 12 \text{ tons}$$
$$6x = 12 \text{ tons}$$
$$x = 2 \text{ tons}$$
$$2x = 4 \text{ tons}$$
$$3x = 6 \text{ tons}$$

     Answer: 2 tons of cement, 4 tons of sand, and 6 tons of gravel

    (b)   If the cost of 9 feet of weatherstripping is $1.50, how much will 72 feet cost?

        Let x = cost of 72 feet

        We are given the ratio 9 feet:$1.50 and must find the equivalent ratio 72 feet:x. The corresponding proportion is

$$\frac{9}{1.50} = \frac{72}{x} \quad \text{or } 9x = 108.00$$
$$x = 12.00$$

     Answer: 72 feet will cost $12.00.

---

* Optional section.

Exercise 51

1. A party punch calls for pineapple juice, orange juice, and lemon juice in a ratio of 8:6:1. How many cups of each must be used to make 240 cups?

2. A 21-foot board is to be cut into three parts in a ratio of 4:2:1. How long is the middle-size piece?

3. Divide $9000 into five parts in a ratio of 5:4:3:2:1.

Use proportions to solve each problem:

4. If 32 square feet of plywood cost $5, what will 100 square feet cost?

5. A dog consumes 28¢ worth of food each day. For how long can he be fed on $10 worth of food?

6. One foot is approximately 30 centimeters. How many centimeters are there in 9 inches?

7. On a map, the scale of one inch equals 3½ miles is used. The distance between two towns is known to be 56 miles. How far apart will the towns be on the map?

8. The water-to-alcohol ratio of a hospital solution is 5:2. How much solution must be mixed if it is to contain 84 gallons of alcohol?

9. A boy can walk 2 miles in the same time it takes him to ride his bike 19 miles. If he rides his bike 95 miles, how far could he have walked in the same amount of time?

10. If 3 ladies spend $146 on clothing, how much would 7 ladies spend at the same rate?

11. Lemons are 6 for 85¢:
    (a) How many can you buy for $6.40?
    (b) How much will 54 lemons cost?

12. Without changing its speed, a car covers 1320 feet in 13.2 seconds. How far can it go in 100 seconds?

Cumulative Review: Chapters 1 through 7

1. Describe the empty set by giving a set which has no members.

2. *Find S:*   (a) $2x - 3 = {}^-3$ (b) $\frac{1}{2}a + 3 = a + 7$

3. *Reduce:* (a) $\dfrac{3m^2 + 17m + 10}{5m^2 - 125}$ (b) $\dfrac{4a^2 + a}{12a^2 + 3a}$

4. *Simplify:* (a) $\dfrac{\dfrac{1}{x} - 2}{x + \dfrac{1}{x}}$ (b) $\dfrac{\dfrac{a}{b} + 1}{\dfrac{a^2}{b^2} - 1}$

5. *Give an example of*
   (a) a linear equation in one variable
   (b) a quadratic trinomial
   (c) a difference of two squares
   (d) a trinomial which is a perfect square
   (e) a real number and its additive inverse

6. *Solve for b:* $k = mgb$

7. *Add:* (a) $\dfrac{3}{4} - \dfrac{1}{6a}$ (b) $\dfrac{3x + 1}{2} - \dfrac{x - 1}{3} + \dfrac{3x + 1}{3x}$

8. *Find S:* $\dfrac{3x + 2}{5} + \dfrac{2x - 9}{3} = 2$

9. *Find S:* (a) $x^2 + 5x + 4 = 0$ (b) $x^2 - 3x = 0$

10. *Factor completely:*
    (a) $3q^2 - 6q$
    (b) $a^2 x^2 - 2ax + 1$
    (c) $16m^2 - 25$

11. State the associative axiom for addition.

12. Find the LCM for $\{3x^2, x^2 - 9, 3x^2 - 9x\}$.

13. What specific type of real number is:
    (a) 6   (b) $^-3$ (c) $\sqrt{3}$ (d) $^-1 \cdot 6$
    (e) $^5\!/_6$ (f) $\sqrt{25}$

14. *Divide* $a^2 + 6a + 8$ by $a + 2$

15. *Simplify:* $^-3(x + 2) - ^-3(x - 8) - 4 + 2a - 6 - ^-5$

16. *Subtract* $^-4m^2 + 2m - 8$ from $16m^3 + 3m^2 + 6$

17. What is to be found in a solution set?

18. Find 5 consecutive even integers whose sum is 230.

19. *Use Proportion:* The sale price for ice cream bon-bons is 12 for 98¢. How many can be purchased for $7.84?

20. Greg has $6.90 in change consisting of nickels, quarters, and half-dollars only. If there are 6 fewer quarters than nickels and only ½ as many half-dollars as quarters, what is the value of all the nickels and quarters combined?

8 Graphs and systems of linear equations

8-1. LINEAR EQUATIONS IN TWO VARIABLES

At this point we have studied linear equations and some quadratic equations in one variable. Our attention will now turn to linear equations in two variables, such as $x + y = 6$. Again we shall attempt to make a conditional equation a true equation and consequently find the solution set. Note that for the equation $x + y = 6$, if $x = 2$ and $y = 4$, then the equation is true. However, if $x = 3$ and $y = 3$ or $x = 5$ and $y = 1$, the equation is still true. In fact, we could go on forever giving roots to this equation. We now give a somewhat formal way of writing the solution set for a linear equation in two variables.

8-2. ORDERED PAIRS

DEFINITION: *(x,y) is called an* ordered pair. *x is the* first member *and y is the* second member *of the ordered pair.*

Consider a set of ordered pairs, $\{(3,3), (2,1), (1,2)\}$. This set contains three members only. The ordered pairs $(2,1)$ and $(1,2)$ are not the same.

To generate an infinite set of ordered pairs, we may use set builder notation:

$$A = \{(x,y) \mid x + y = 6\}$$

Set A is a set of ordered pairs such that the sum of the two members of each ordered pair is 6.

8-3.　THE RECTANGULAR COORDINATE SYSTEM

All of the ordered pairs used in this text will be pairs of real numbers. It will be most convenient for us to have a method of picturing all of the ordered pairs. The French philosopher and mathematician René Descartes (1596-1650) invented a method for doing this, which is called the *Cartesian coordinate system* in his honor, or more generally, the rectangular coordinate system (see Figure 8-1). Construct two lines which intersect perpendicular to each other. Choose the point of intersection, called the *origin*, as a starting point, and mark the horizontal axis just as we did for the real number line in chapter 1. Call the *horizontal axis the x-axis* or the *axis of abscissas*. On the vertical axis make divisions similar to those on the x-axis, with the portion above the x-axis for positive real numbers and the portion below for the negative numbers. This *vertical axis* is called the *y-axis* or the *axis of ordinates*.

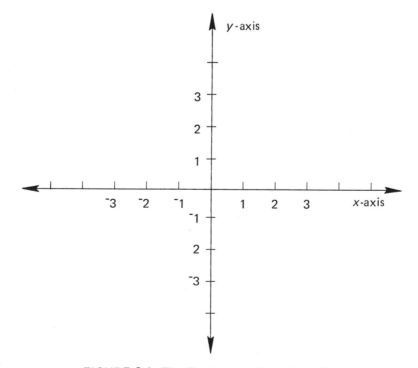

FIGURE 8-1　The Rectangular Coordinate System

This coordinate system, which lies in a plane and is two dimensional, can be used for plotting ordered pairs of real numbers. Each point in the plane will be given the name of a unique ordered pair. For example, let us begin a the origin and move along the x-axis two units to the right, then move parallel to the y-axis three units in a positive (upward) direction. The point at which we stop shall be called (2,3). In general, the first member gives the movement in the x-direction. and the second member gives the movement in the y-direction. If the point (2,3) is denoted as point P, the we say P has an x-coordinate and a y-coordinate of 2 and 3, respectively. We may write this as P(2,3). (See Figure 8-2.)

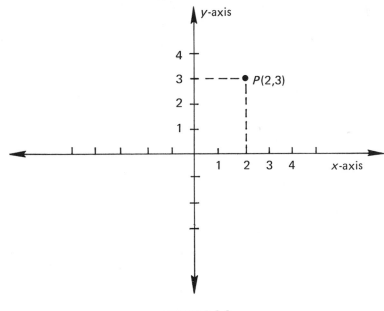

FIGURE 8-2

DEFINITION: *To* plot a point *(x,y) means to find the point in the plane whose coordinates are x and y.*

DEFINITION: *The x-coordinate of a point is called the* abscissa. *The y-coordinate of a point is called the* ordinate.

example:

Study the following graph (picture) of several points to see how they are plotted.

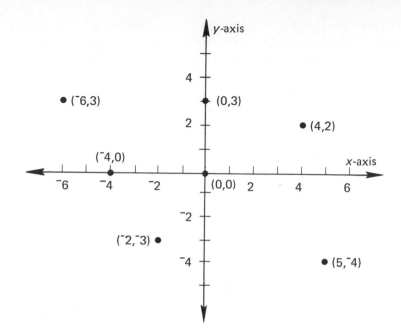

The pair of coordinate axes divide the plane into four quarter planes called *quadrants*. These quadrants are numbered in a counter-clockwise direction beginning with the upper right quadrant. (See Figure 8-3.)

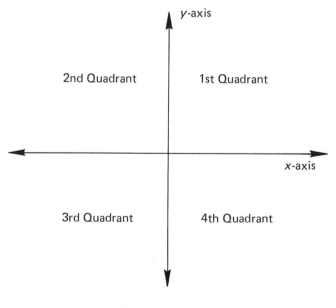

FIGURE 8-3

examples: Given: $A = \{(x,y) \mid x - 2y = 6\}$

(a)   Find y if $(8,y) \in A$:

This means that $x = 8$ and $x - 2y = 6$ must
be made true.

Substituting: $8 - 2y = 6$

$$^-2y = {}^-2$$

$$y = 1$$

Answer: When $y = 1$, $(8,y) \in A$.

(b)   Find x if $(x,3) \in A$:

This means that $y = 3$ and $x - 2y = 6$
must be made true.

Substituting: $x - 2(3) = 6$

$$x - 6 = 6$$

$$x = 12$$

Answer: When $x = 12$, $(x,3) \in A$.

Exercise 52

1. Given the set of ordered pairs $A = \{(x,y) \mid x + y = 3\}$:
    (a)   find 5 ordered pairs $(x,y) \in A$.
    (b)   is $(3,^-6) \in A$?
    (c)   is $(28,^-25) \in A$?
    (d)   find y if $(3,y) \in A$.
    (e)   find y if $(^-4,y) \in A$.
    (f)   find x if $(x,^-2) \in A$.
    (g)   find x if $(x,6) \in A$.

2. Given $B = \{(x,y) \mid x - y = 8\}$, answer the same questions
    (a) through (g) from problem 1.

3. Set up a rectangular coordinate system, label the two coordi-
    nate axes, and plot the following points:
    (a) (2,2)   (b) (6,3)   (c) (1,5)   (d) ($^-$1,2)
    (e) ($^-$1,$^-$3)   (f) ($^-$3,4)   (g) (5,$^-$2)   (h) (0,5)
    (i) (5,0)   (j) ($^-$6,0)

4. Construct a rectangle by plotting $P(3,4)$, $Q(3,^-6)$, $R(^-2,4)$,
    and $S(^-2,^-6)$.

5. *Find S:*   (a) $x + \dfrac{1}{2} = 3x - 4$ (b) $x(x - 2) = 0$

             (c) $2b - (3 - b) = 6$ (d) $\dfrac{a + 1}{2} + a = 4$

6. *Reduce:* (a) $\dfrac{2t^2 - t}{2t^2 + t}$ (b) $\dfrac{x + 1}{x^2 - 1}$ (c) $\dfrac{3y^2}{y^2 - y^3}$

7. *Add:*   (a) $\dfrac{1}{x} + 2$ (b) $\dfrac{3}{4} + \dfrac{1}{a}$ (c) $\dfrac{a^2x^3}{y} + \dfrac{a^2y^2}{x}$

8. In a rectangular coordinate system, the horizontal axis is called the _____ or _____ .

9. If two points have the same x coordinate, say (3,5) and (3,8), are they the same distance from: (a) the origin? (b) the y-axis?

10. If two points have the same y-coordinate, then they are the same distance from _____ . (Make a drawing to illustrate your answer.)

11. *Factor:*   (a) $a^2 + 6a - 7$ (b) $2ax^2 - 4ax + 2a$

12. *Multiply:*   (a) $(3x + 2)(3x - 2)$ (b) $3x(x + 2)^2$
                (c) $(2m - 1)(5m + 6)$ (d) $(a + 3)^2 (a + 3)$

8-4. THE GRAPH OF A LINEAR EQUATION IN TWO VARIABLES

We have seen that the solution set for a linear equation, such as $x + y = 6$, contains an infinite number of ordered pairs. If these ordered pairs are all plotted on the same coordinate system, they will form a straight line. The straight line is the graph of the solution set.

THEOREM 8-1: *The graph of a linear equation in two variables is always a straight line.*

(This theorem can be proven in a more advanced course called analytic geometry.)

We now borrow an axiom from geometry:

AXIOM: *Through two distinct points one and only one straight line may be drawn.*

Combining the two ideas in the above theorem and axiom allows us to graph the set $B = \{(x,y) \mid x + y = 4\}$ by merely finding two points in B, plotting these two points and then drawing the straight line which passes *through* two points.

examples: (a) Graph $D = \{(x,y) \mid x + y = 7\}$

First, find two members of D:

If $x = 1,$* then $x + y = 7$ becomes $1 + y = 7$ or $y = 6$, which gives $(1,6)\ e\ D$.

If $x = 3$, then $x + y = 7$ becomes $3 + y = 7$ or $y = 4$ and $(3,4)\ \epsilon\ D$.

As a check of the accuracy of our two points, let $x = 5$, so that $5 + y = 7$ or $y = 2$ and $(5,2)\ \epsilon\ D$. This point should lie on the same line as the first two points.

Second, plot the three points, $(1,6)$, $(3,4)$, and $(5,2)$ on the same pair of axes.

Third, pass a line through the three points and write the name of the line, $x + y = 7$ along the line.

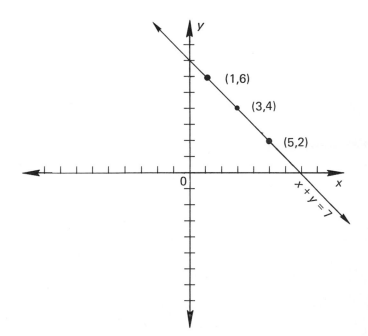

---

*The choice of $x = 1$ is completely arbitrary. Any value for x or y may be selected and the corresponding y or x can be found by direct substitution. We usually choose numbers which are small and convenient to use.

(b) Graph $A = \{(x,y) \mid x = 2\}$

Members of this set are very easy to find, as each one has $x = 2$ and y can be any real number.

Three such members are (2,3), (2,4), and (2,0). When these three points are plotted, we get a line parallel to the y-axis and two units to the right.

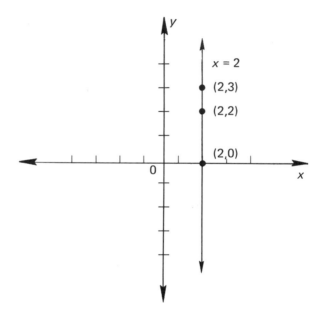

Exercise 53

Graph each set:

1. $\{(x,y) \mid x + y = 5\}$ 2. $\{(x,y) \mid x - y = 2\}$
3. $\{(x,y) \mid x + y = 10\}$ 4. $\{(x,y) \mid x + 2y = 4\}$
5. $\{(x,y) \mid 2x - y = 1\}$ 6. $\{(x,y) \mid y = x\}$
7. $\{(x,y) \mid y = 2x\}$ 8. $\{(x,y) \mid y = x + 3\}$
9. $\{(x,y) \mid 3x + 2y = 6\}$ 10. $\{(x,y) \mid x = 3 - y\}$
11. $\{(x,y) \mid 3y = 2x + 1\}$ 12. $\{(x,y) \mid y = \frac{1}{2}x + 3\}$
13. Graph the line $2y + 3x = 18$ by plotting the points (0,y) and (x,0) which satisfy the equation.
14. Graph $x - 3y = 6$ using the points (3,y) and (x,1).
15. Is $(3,4) \in \{(x,y) \mid 2x - y = 2\}$? Show why it is or is not.
16. Name five prime numbers between 20 and 50.
17. Graph the line $y = 3x - 1$.

18. Graph the two lines $y = 3$ and $x = {}^-2$.
19. On the same pair of axes, graph:
    (a) $\{(x,y) \mid x - y = 1\} \cap \{(x,y) \mid x + y = 5\}$
    (b) $\{(x,y) \mid x = 8\} \cap \{(x,y) \mid 2x + y = 4\}$
    (c) $\{(x,y) \mid y = 3\} \cap \{(x,y) \mid y = 3x\}$
    (d) $\{(x,y) \mid 2x + 4y = 10\} \cap \{(x,y) \mid x - y = 4\}$
20. What can be said about two straight lines that intersect more than once?
21. Is $3 \in \{(x,y) \mid x = 3\}$? Explain.
22. Is $(0,3) \in \{(x,y) \mid x = 3\}$? Explain.
23. Is $(0,0) \in \{(x,y) \mid y = x\}$? Name three members of this set.
24. Graph $\{(x,y) \mid 5x - 3 = 2y\}$
25. Graph $8 = 3x - 6 + 2y$

8-5. INTERCEPT METHOD OF GRAPHING

With the exception of those lines which are parallel to an axis, all straight lines must pass through the x- and y-axes.

DEFINITION: *The x-coordinate of the point on the x-axis through which a line passes is called the x-intercept of the line.*

DEFINITION: *The y-coordinate of the point on the y-axis through which a line passes is called the y-intercept of the line.*

We know that every point on the x-axis has a y-coordinate of zero. Therefore, the *x-intercept of a straight line occurs at the point whose y-coordinate is zero.* The y-intercept occurs when $x = 0$.

Since we need only two points to determine a line, the two intercepts are very convenient to use when plotting a line.

examples: (a) Graph $\{(x,y) \mid 2x - 5y = 20\}$ using the intercept method.

If $y = 0$, then the point $(x,0)$ contains the x-intercept.

The equation $2x - 5y = 20$ becomes
$$2x - 5(0) = 20$$
$$2x = 20$$
$$x = 10, \text{ the } x\text{-intercept}$$

To find the y-intercept, let $x = 0$:

$$2x - 5y = 20 \quad \text{becomes}$$
$$2(0) - 5y = 20$$
$$^{-}5y = 20$$
$$y = {}^{-}4, \text{ the } y\text{-intercept}$$

Using the two points that contain the intercepts, we plot $(10,0)$ and $(0,{}^{-}4)$ and thus have determined the line.

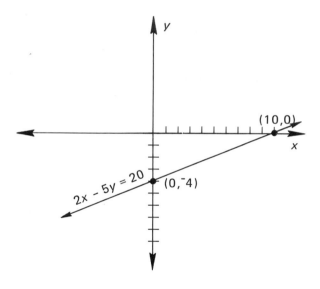

(b) Graph $\{(x,y) \mid y = 3x\}$

To find the y-intercept, let $x = 0$

$$y = 3(0)$$
$$y = 0$$

This tells us that $(0,0)$ contains both intercepts and we must find another point in order to plot the line. Choose any x, say $x = 2$.

$$y = 3(2)$$
$$y = 6$$

This makes $(2,6)$ a point on the line $y = 3x$. Plotting these two points, we get the following graph:

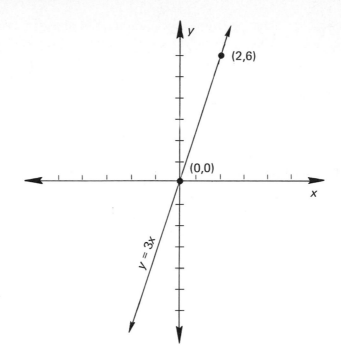

Exercise 54

Use the intercept method to graph the solution set for each equation:

1. $\{(x,y) \mid x + y = 5\}$
2. $\{(x,y) \mid x - y = 2\}$
3. $\{(x,y) \mid x + y = {}^-4\}$
4. $\{(x,y) \mid x - 2y = 6\}$
5. $\{(x,y) \mid 3x + y = 6\}$
6. $\{(x,y) \mid x = 2y + 8\}$
7. $\{(x,y) \mid y = 2x\}$
8. $\{(x,y) \mid y = x - 3\}$
9. $\{(x,y) \mid 3y = x + 12\}$
10. $\{(x,y) \mid 5x = 2y + 10\}$

Graph the next five lines (problems 11 through 15) by plotting two points in the solution set, and then find the intercepts as a means of checking:

11. $2y + 3y = 12$
12. $y = 6x + 10$
13. $5y - x = 15$
14. $3y - 2x = 14$
15. $\dfrac{1}{2}x + \dfrac{1}{3}y = 1$

16. If $(3,y) \in A = \{(x,y) \mid 2x + y = 10\}$, find y.
17. If $(0,y) \in B = \{(x,y) \mid x - y = {}^-3\}$, find y.
18. If $(x, {}^-2) \in C = \{(x,y) \mid 5 = 3x - 2y\}$, find x.

19. What are the x- and y-intercepts of $2y - x = 6$?

20. What is the y-intercept of $\{(x,y) \mid x = 3\}$?

21. *Graph:* $\{(x,y) \mid x = {}^-9\}$

22. *Graph:* $\{(x,y) \mid y = 2\}$

23. *Find S:* $3(x + 2) - 4(2x - 3) = {}^-2$

24. *Find S:* $\dfrac{2x - 5}{3} + 3 = \dfrac{1}{2}$

25. *Reduce:* (a) $\dfrac{x^2 - 3x}{x^2 - 9}$ (b) $\dfrac{a + b}{a^2 - b^2}$ (c) $\dfrac{4x - 3}{20x - 5}$

26. *Add:* (a) $\dfrac{3x}{5} + \dfrac{2}{3x}$ (b) $\dfrac{x + 2}{x} + \dfrac{1}{x + 1}$

27. *Multiply:* $\dfrac{2x - 6}{x + 2} \cdot \dfrac{3x + 6}{3x^2 - 9x}$

28. *Find the intercepts and graph:* $x = 3y - 2$

29. Find three ordered pairs that are members of $\{(x,y) \mid 5x + 2y = {}^-8\}$.

30. Find x and y if $(x,0)$ and $(0,y)$ are members of $\{(x,y) \mid x = y - 4\}$.

8-6. SOLVING A SYSTEM OF TWO LINEAR EQUATIONS BY GRAPHING

If we are looking for two numbers whose sum is 12 and whose difference is 4, we may let x and y represent the two unknown numbers and write

$$(1) \quad x + y = 12$$
$$(2) \quad x - y = 4$$

Equations (1) and (2) are called a *system of equations, since a solution common to both equations is sought.*

To solve a system of equations we must find the intersection of the solution sets of each equation in the system. For equations (1) and (2) above, $S = \{(x,y) \mid x + y = 12\} \cap \{(x,y) \mid x - y = 4\}$, which we may observe to be $\{(8,4)\}$. The algebraic methods for finding the common solution will be discussed in the next two sections.

The *graph of a system of two linear equations in two variables is two straight lines.* The ordered pair(s) in the solution set will appear as the point(s) of intersection of the two lines. If there is exactly one point of intersection, then the system is said to be *linearly independ-ent* or *simultaneous*. If the two lines are parallel, which means that the solution set is empty, then the system is *inconsistent*. Finally, if the two lines intersect more than once, then the two lines are the same line. The solution set in this case contains all the points on that line and is an infinite set. This system is said to be *dependent*.

A system of equations can be solved by graphing each straight line and simply reading the coordinates of the point of intersection of the lines. However, this method is not too accurate, especially if the point of intersection has fractional coordinates; but it does give good approx-imate roots. (You must use graph paper for good results.)

examples: (a)   Find S by graphing the system

$\{(x,y) \mid x + 3y = 11\} \cap \{(x,y) \mid 4x - y = 5\}$

First, graph each line, using the methods of section 8-4 or 8-5.

(1) $x + 3y = 11$

(2) $4x - y = 5$

Second, extend the lines beyond the point of intersection and read the coordinates of this point.

Answer: $S = \{(2,3)\}$

Note: S contains only *one* member.

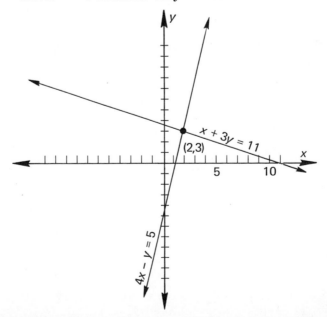

(b)   Find S by graphing: (1) $x + y = 6$
                              (2) $x + y = 2$

It seems that these two lines are parallel.
Therefore, $S = \emptyset$.

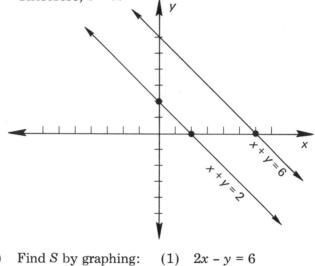

(c)   Find S by graphing: (1) $2x - y = 6$
                              (2) $^-4x + 2y = ^-12$

If we use the intercept methods, we find that
each line has the same x- and y-intercepts, 3 and
$^-6$, respectively.

Answer: $S = \{(x,y) \mid 2x - y = 6\}$

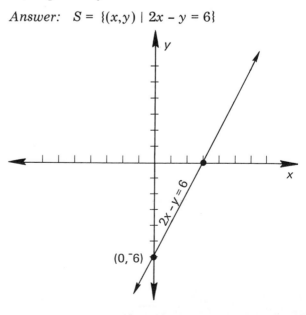

Exercise 55

For each system of equations, find S by graphing:

1. (1) $x + y = 5$
   (2) $x - y = {}^-1$

2. (1) $3x + y = 7$
   (2) $x - 3y = 7$

3. (1) $4x - 5y = {}^-7$
   (2) ${}^-2x + 4y = 2$

4. (1) $y - x = 4$
   (2) $2y + 6x = {}^-4$

5. (1) $y = 2x$
   (2) $y = 6x$

6. (1) $y = x + 3$
   (2) $y = x - 4$

7. (1) $3x - y = 2$
   (2) $2y = 6x + 4$

8. (1) $5x - y = 3$
   (2) ${}^-5x + y = {}^-3$

9. (1) $x = 2$
   (2) $y = 3$

10. (1) $x + 3 = {}^-5$
    (2) $2y = 6$

11. If the graph of a system of equations shows two parallel lines, then the system is called _____ .

12. *Name each axiom shown:*
    (a) $a + b = b + a$
    (b) $x(y + z) = xy + xz$
    (c) $a(bc) = a(cb)$
    (d) $(ab)c = a(bc)$

13. Find the additive inverse of ${}^-3$.

14. *Graph:* $S = \{(x,y) \mid 2x - y = 3\} \cap \{(x,y) \mid x + 3y = {}^-2\}$

15. Find x- and y-intercepts for $8x + 3y = 18$.

16. Find three members of $\{(x,y) \mid 4x - 7y = 6\}$.

17. Find three members of $\{(x,y) \mid y = 8x\}$.

18. Is $(0,3) \in \{(x,y) \mid x + 3 = 6\}$?

19. Show why $(3,3) \notin \{(x,y) \mid y = 5x - 2\}$.

20. *Factor:* (a) $b^2 - 3b$ (b) $b^2 - 3b + 2$ (c) $b^2 - 4$

21. *Find S:* (a) $(x - 2)(2x + 3) = 0$ (b) $x^2 - x = 0$

22. *Find S:* $2y - 3(y + 4) = y - 5y + 2(3 + 4y)$

23. *Solve by graphing:* (1) ${}^-x + 2y = {}^-2$
                          (2) $3x - 4y = 8$

24. *Distinguish between these two sets:* $\{x \mid x = 3\}$ and $\{(x,y) \mid x = 3\}$.

25. Name five irrational numbers.

26. *Subtract* $3x - 2y + 7$ from $8x - 3y - 6$.

27. *Graph:* $\{(x,y) \mid x = 2\} \cap \{(x,y) \mid y - x = 5\}$

28. Use the intercept method to graph $4x + 3y = 36$.

29. *Solve for y:* (a) $3y = {}^-2y + k$ (b) $a + 4 = \dfrac{1}{2}y - a^2$

30. A collection of dimes, nickels, and quarters has the same number of each type of coin. How many coins are there if the
value of the collection is $4.40?

8-7. ADDITION METHOD OF SOLVING A SYSTEM OF EQUATIONS

The addition axiom for equality, introduced in chapter 4, allows
us to add equal quantities to each side of an equation without affect-
ing the equality. As an application of this axiom, consider the following treatment of a system of equations:

$$(1) \quad x + y = 7$$
$$(2) \quad x - y = 1$$

Add the left-hand member of equation (2) to the left-hand member
of equation (1), and then add the right-hand members of both equations to obtain

$$(3) \quad 2x = 8$$
$$(3a) \quad x = 4$$

Equation (3a) tells us that $x = 4$ in equations (1) and (2). Now all
we must do is substitute this value for x in either (1) or (2) and solve
for y. Using (1) we get

$$(4) \quad 4 + y = 7$$
$$(4a) \quad y = 3$$

From equations (3a) and (4a) we conclude $S = \{(4,3)\}$.

The idea behind this addition method is to add two equations in
such a way as to eliminate one of the variables from the resulting
third equation. To accomplish this we must sometimes resort to our
own ingenuity. Study the following examples.

examples: (a)   Find S for   (1) $3x + 2y = 10$
                                    (2) $2x - y = 9$

If (1) and (2) are added, neither variable is
eliminated. But if equation (2) is multiplied
by 2, then we get

(2a) $4x - 2y = 18$ Now add this to
(1) $3x + 2y = 10$ to obtain
(3) $7x = 28$
(3a) $x = 4$

Substitute $x = 4$ into equation (2) to obtain

(4)　$2(4) - y = 9$

(4a)　　$8 - y = 9$

(4b)　　　$^-y = 1$

(4c)　　　$y = {}^-1$

Answer:　$S = \{(4,^-1)\}$

Check:　If $x = 4$ and $y = {}^-1$

(1)　　　$3x + 2y = 10$

$3(4) + 2(^-1) = 10$

$12 - 2 = 10$

$10 = 10$

(2)　　　$2x - y = 9$

$2(4) - (^-1) = 9$

$8 + 1 = 9$

$9 = 9$

(b)　Find members of S if

$S = \{(x,y) \mid 3x + 4y = 11\} \cap \{(x,y) \mid 2x + {}^{\bullet}7y = 29\}$

We are being asked to solve the system

(1)　$3x + 4y = 11$

(2)　$2x + 7y = 29$

The plan is to multiply equations (1) and (2) by appropriate numbers to obtain coefficients of x which are negatives of each other.

Accordingly, multiply equation (1) by 2 and equation (2) by $^-3$, giving

(1a)　　$6x + 8y = 22$

(2a)　$^-6x - 21y = {}^-87$

Add (1a) to (2a);

(3)　　　$^-13y = {}^-65$

(3a)　　　$y = 5$

Substitute (3a) into (1):

(4)　$3x + 4(5) = 11$

(4a)　$3x + 20 = 11$

(4b)　　　$3x = {}^-9$

(4c)　　　$x = {}^-3$

Answer:　$S = \{(^-3,5)\}$

Check: If $x = {}^-3$ and $y = 5$

(1) $3x + 4y = 11$

$3({}^-3) + 4(5) = 11$

${}^-9 + 20 = 11$

$11 = 11$

(2) $2x + 7y = 29$

$2({}^-3) + 7(5) = 29$

${}^-6 + 35 = 29$

$29 = 29$

Exercise 56

Find S without graphing the system:

1. (1) $x = 3$
    (2) $3x - 2y = 11$

2. (1) $y = {}^-2$
    (2) $x + y = 12$

3. (1) $3x - 8y = {}^-12$
    (2) $3x = 12$

4. (1) $3y = 18$
    (2) $3x = 40$

5. (1) $x = 6$
    (2) $y = {}^-14$

6. (1) $\frac{1}{2}x = \frac{7}{5}$

    (2) $\frac{3}{5}y = \frac{1}{3}$

Use the method of addition to find S:

7. (1) $x + y = {}^-3$
    (2) $x - y = {}^-11$

8. (1) $x - y = 5$
    (2) $2x + y = 7$

9. (1) $6y - x = 4$
    (2) $5y + x = {}^-15$

10. (1) $y = 5 - x$
     (2) $2y = 7 + x$

11. (1) $3x + 2y = 5$
     (2) $5x - 2y = 3$

12. (1) ${}^-4x - 3y = 9$
     (2) $5x + 3y = {}^-3$

13. (1) $x + 2y = 10$
     (2) ${}^-3x + 5y = 14$

14. (1) $6x + 5y = {}^-11$
     (2) $4x - y = {}^-3$

15. *Square:*   (a) $3x$ (b) $(3x - 2)$
16. *Graph:*   (a) $\{(x,y) \mid x = {}^-1\}$ (b) $\{(x,y) \mid y = 3\}$
17. Find the x- and y-intercepts for $3y = x + 2$.
18. *Solve by addition:*   (1) $2x - 3y = 10$ and (2) $5x + 3y = 4$
19. *Find S:* $\{(x,y) \mid x + 3y = 4$ and $y = 3\}$

Solve by addition:

20. (1) $x - 8 = y$
    (2) $y = 3x - 12$

21. (1) $4x = 3y + 3$
    (2) $3y = 2x - 1$

22. (1) $\dfrac{x}{6} + \dfrac{y}{3} = 1$

    (2) $x - \dfrac{y}{3} = \dfrac{11}{3}$

23. (1) $\dfrac{2x}{3} + y = 2$

    (2) $\dfrac{3x}{5} - y = \dfrac{9}{5}$

24. *Find S:* $\dfrac{x + 3}{2} + \dfrac{3x - 1}{3} = x + \dfrac{1}{3}$

25. $(3x - 7)^2 \neq 9x^2 - 21x + 49$. Why?
26. Find k such that $(3,^-6) \in \{(x,y) \mid 3x - 2y = k\}$.
27. Find k such that $(^-4,1) \in \{(x,y) \mid 5x + 3y = k\}$.
28. Find x such that $(x,^-\frac{1}{2}) \in \{(x,y) \mid 2x - 3y = 5\}$.
29. Distinguish an axiom from a theorem.
30. Name three negative rational numbers.
31. *Find S:* (1) $3x + 2y = 11$
    (2) $x - 3y = 0$
32. *Find S:* (1) $3x - 2y - 6 = 0$
    (2) $4x + 6y + 9 = 0$

8-8. SUBSTITUTION METHOD OF SOLVING A SYSTEM OF EQUATIONS

The substitution method is now illustrated. To solve the system (1) $x + y = 4$, (2) $y = x + 3$, observe that equation (2) is solved for y in terms of x. Since the x's and y's in each equation represent the same respective numbers, the expression $y = x + 3$ from (2) can be substituted into (1). Equation (1) becomes (3) $x + (x + 3) = 4$.

To summarize: Find S if

$$(1) \quad x + y = 4$$
$$(2) \quad y = x + 3$$

Substitute (2) into (1):

$$(3) \quad x + (x + 3) = 4$$
$$(3a) \quad 2x + 3 = 4$$
$$(3b) \quad 2x = 1$$
$$(3c) \quad x = \tfrac{1}{2}$$

Substitute (3c) into (2):

$$\text{(4)} \qquad y = \tfrac{1}{2} + 3$$
$$\text{(4a)} \qquad y = 3\tfrac{1}{2}$$

Answer: $\quad S = \{(\tfrac{1}{2}, 3\tfrac{1}{2})\}$

example: Find S by substitution for

$$\text{(1)} \qquad 2x - 3y = 12$$
$$\text{(2)} \qquad 3x - y = 11$$

We must first solve one of the equations for one of the variables. Solving (2) for y seems most convenient.

$$\text{(2a)} \qquad y = 3x - 11$$

Substitute (2a) into (1):

$$\text{(3)} \quad 2x - 3(3x - 11) = 12$$
$$\text{(3a)} \quad 2x - 9x + 33 = 12$$
$$\text{(3b)} \qquad {}^-7x = {}^-21$$
$$\text{(3c)} \qquad x = 3$$
$$\text{(4)} \qquad y = 3(3) - 11$$
$$\text{(4a)} \qquad y = {}^-2$$

Answer: $\quad S = \{(3, {}^-2)\}$

Exercise 57

Use the method of substitution to find S:

1. (1) $2x - y = 6$
   (2) $y = x - 2$

2. (1) $x + 2y = 5$
   (2) $y = 3 - x$

3. (1) $x = 8$
   (2) $5x - 3y = 1$

4. (1) $y = 18 - {}^-3$
   (2) $2x - 4 = y$

5. (1) $x + y = 3$
   (2) $x - 3y = {}^-1$

6. (1) $x = 2y - 4$
   (2) $3x - y = 1$

7. (1) $y = \tfrac{1}{2}x - 3$
   (2) $4y - 5x = 20$

8. (1) $\tfrac{1}{2}x = 4 + y$
   (2) $3x - 15y = 15$

9. (1) $y = 3x - 4$
   (2) $y = x + 6$

10. (1) $x - 3 = 3y$
    (2) $x = y$

11. (1) $y = \dfrac{x + 3}{4}$
    (2) $2x + 5y = 46$

12. (1) $4x = {}^-y - 1$
    (2) $x = \dfrac{3 - 4y}{9}$

13. *Find S:* $(x - 4)(x + 2) = 0$
14. *Find S:* $4x(x - 8) = 0$
15. *Find S by addition:*   (a) (1) $3x + 2y = 11$
    (2) $5x - 2y = 5$
    (b) (1) $x + \frac{1}{2}y = 3$
    (2) $x - \frac{1}{6}y = {}^-1$
16. If a system of equations is inconsistent, then the solution
    set contains _____ and the graph shows
    _____ lines.

Use substitution to find S:

17. $\{(x,y) \mid (1)\ x + y = 24$ and $(2)\ x = 2y + 42\}$
18. $\{(x,y) \mid y = x + 3\} \cap \{(x,y) \mid {}^-3x + 5 = 2y + 3\}$
19. (1) $y = 2a + 1 + x$ 20. (1) $2x - y = a$
     (2) $x - 3y = a$ (2) $5x - 7y = 0$
21. (1) $y = 3x$ 22. (1) $y - 3 = \frac{1}{2}y - \frac{1}{3}$
     (2) $4x - 6y = 1$ (2) $x + y = 12$
23. State the addition axiom for equality (see section 4-3).
24. *Factor:*   (a) $ax + ay$ (b) $2ax^2 - 8a$
                 (c) $6x^2 + 19x - 7$ (d) $m^3 - 6m^2 + 9m$

25. *Add:*   (a) $\dfrac{1}{x} + 4$ (b) $\dfrac{2}{a + b} - \dfrac{3}{a - b}$

              (c) $x + \dfrac{1}{x} + \dfrac{1}{x^2}$ (d) $9ax + \dfrac{3}{x} + \dfrac{ax^2}{x}$

26. *Find S:*   (1) $x - 2y = 13$ and (2) $3x + 2y = {}^-9$
27. *Graph:* $\{(x,y) \mid x = 3\} \cap \{(x,y) \mid y = {}^-1\}$
28. *Give an example of*
     (a) a linear equation in two variables
     (b) the product of the sum and difference of two terms
     (c) a description of the empty set
     (d) a second-degree binomial that is prime
     (e) a number and its multiplicative inverse
     (f) the additive identity
29. *Find S:*   (1) $y - 3 = 2x + 7$
                 (2) $y = 4 - 3x$

30. *Find S:*   (1) $\dfrac{x}{3} + \dfrac{y}{5} = 6$

                 (2) $\dfrac{2x}{5} + \dfrac{3y}{4} = 1$

Exercise 58

Find S by any method except graphing:

1. (1) $2y + 4x = {}^-14$
   (2) $5y - x = {}^-2$

2. (1) $x = {}^-\frac{1}{3}$
   (2) $y = {}^-\frac{1}{2}$

3. (1) $x - 3y = 5$
   (2) $3x - y = 0$

4. (1) $2x - 5y = 1$
   (2) $y = 2x$

5. (1) $6x + 5y = 10$
   (2) $4x - 2y = 5$

6. (1) $5x - 7y = {}^-2$
   (2) $6x + 2y = 8$

7. (1) $4x - 4y = 4$
   (2) $6x + 6y = {}^-6$

8. (1) $3x - 5y = 10$
   (2) $4x + 3y = {}^-6$

9. (1) $a_1 x + b_1 y = c_1$
   (2) $a_2 x + b_2 y = c_2$

10. (1) $kx - 3y = 2$
    (2) $3kx - 7y = 1$

11. $3(x - 2) + 4x - 2(3 - x) = 2 - 13$

12. $x^2 - 5x - 6 = 0$

13. $x^2 - 3x = 0$

14. $\{(x,y) \mid 2x - y = 1\} \cap \{(x,y) \mid x + y = 0\}$

15. (1) $0 = x + y - 6$
    (2) $0 = 5x + 2y + 8$

8-9. STATED PROBLEMS USING TWO EQUATIONS AND TWO VARIABLES

If a stated problem calls for two answers, it is very natural to use a different variable for each one. The use of two variables requires that we write two equations involving the variables and solve the resulting system of equations.

examples: (a) Find two numbers whose sum is 13 and whose difference is 5.

Solution: Let x be the first number and y be the second number.

Then
(1) $x + y = 13$
(2) $x - y = 5$

Solve by addition:
(3) $2x = 18$
(3a) $x = 9$
(4) $9 + y = 13$
(4a) $y = 4$

From (3a) and (4a) the numbers are 9 and 4.

(b)   Helen has 18 coins consisting of dimes and nickels. Their value is $1.25. How many of each does she have?

Solution: Let d be the number of dimes and n be the number of nickels.

Then   (1) $d + n = 18$

              (2) $10d + 5n = 125$

                         (penny value equation)

Solve by sub-
stitution: (1a) $n = 18 - d$

              (3) $10d + 5(18 - d) = 125$

              (3a) $10d + 90 - 5d = 125$

              (3b) $5d = 35$

              (3c) $d = 7$

              (4) $7 + n = 18$

              (4a) $n = 11$

From (3c) and (4a) we get $d = 7$ dimes and $n = 11$ nickels.

Exercise 59

For each problem use two variables:

1. Find two numbers whose sum is 90 and whose difference is 40.
2. Find two numbers whose sum is ‾35 and whose difference is 45.
3. Find two positive integers whose product is 28 and where the first number is 3 more than the second.
4. If one number is 3 times another and their sum is 24, find the numbers.
5. A car and boat cost a total of $5600. The car cost $3000 more than the boat. How much did each cost?
6. A house and lot cost $15,600. The house is valued at 4 times the lot. What is the cost of each?
7. A rectangular garden has a perimeter of 300 feet, and the length is 40 feet greater than the width. What are the dimensions of the garden?
8. The length of a rectangle is twice the width, and the perimeter is 282 feet. What is the length?
9. A 30-foot board is cut into two parts so that one part is 2½ times the other. How long is the smaller piece?

10. A man receives $463 income from two accounts. One account pays $46 more than the other. How much does he receive from each?

11. During 1972, rainfall increased by .3 inch over the 1971 rainfall. If the total for the two years was 14.8 inches, how much rain fell each year?

12. Students paid 40¢ for theater tickets and adults paid 90¢. If 360 tickets were taken and $204 received, how many adults attended?

13. A change box contains 38 nickels and twice as many quarters as dimes. After counting all this change, the value is set at $121.90. How many dimes are there?

14. During a supermarket sale, cans of corn are priced at 19¢ and cans of pineapple at 24¢. A lady buys 10 more cans of pineapple than corn and has to pay $5.84. How much did she pay for the corn?

15. Two boys are sitting at the ends of a 12-foot teeter-totter that is balanced. If one boy weighs 80 pounds and the other 100 pounds, how far from the heavier boy is the fulcrum? (Make a sketch.)

16. A total weight of 600 pounds is to be distributed at each end of a lever and fulcrum so as to balance the lever. If the fulcrum is 8 feet from one end and 10 feet from the other, how much weight must be placed at each end?

17. Sam has $6000 to invest. Some of his money returns 7½% and the rest returns 6%. Over a period of one year he receives $414 interest. How much is invested at each rate?

18. The amounts of interest received from two savings accounts are equal. The amount of money invested is $640. How much was invested in each account if one pays 7% and the other 6%?

19. To finance a trip, a man must sell his diamond ring and his wife's fur coat. He receives $3200 for both the ring and coat, and the ring sold for ⅔ as much as the coat. How much did he get for the coat?

20. Find two numbers whose difference is 8 and where the difference of their two squares is 128.

Cumulative Review: Chapters 1 through 8

1. Draw the real number line and place on it the numbers $^-3$, ½, 0, π, $^-1\frac{7}{3}$, 6.3, 6.06, $^-4$, $^-3.8$, and $^-5.5$.

2. *Subtract* $3x^2 - 4x - 9$ from $5x^3 + 2x^2 - 3x + 7$.

3. *Divide:* $x + 3 \overline{\smash{\big)}\ 3x^3 + 2x^2 - 3x + 16}$

4. *Simplify:* (a) $\dfrac{\dfrac{4}{x} + 2}{\dfrac{8}{x} + \dfrac{2}{3}}$ (b) $\dfrac{1}{1 - \dfrac{1}{x}}$

5. *Find S:* (a) $3x - 4 = 11$ (b) $\frac{3}{4}x = 5$
    (c) $3x^2 - 4x = 0$ (d) $x^2 - 4x - 5 = 0$

6. *Graph:* (a) $\{(x,y) \mid x - y = 5\}$ (b) $\{(x,y) \mid x = y\}$
    (c) $\{(x,y) \mid x = 2 \text{ and } y = 3\}$ (d) $\{(x,y) \mid y = {}^-3\}$

7. *Find S:* (1) $2x - y = 4$
    (2) $3x + 2y = 27$

8. *Factor completely:* (a) $mx^2 - 4m$ (b) $a^2 + 12a + 36$
    (c) $24x^2 - 150$ (d) $4x^2 + 16$

9. *Add:* (a) $\dfrac{1}{x} + \dfrac{1}{2}$ (b) $\dfrac{2}{3x} + \dfrac{1}{6}$

    (c) $3 + \dfrac{1}{x}$ (d) $\dfrac{3}{a + b} - \dfrac{2}{a + b}$

10. *Multiply:* (a) ${}^-3(2x - 4)$ (b) $(x + 5)^2$
    (c) $(2x + 7)(x - 3)$ (d) $4x(x + 2)(3x - 1)$

11. Which axiom allows us to remove a common factor?

12. Is $3 \epsilon \{(x,y) \mid x = 3\}$? Why?

13. Is $3 \epsilon \{x \mid x = 3\}$? How many members are in this set?

14. Using two variables, separate a 27-foot piece of rope into two parts such that one part is 3 more than $\frac{1}{2}$ the other.

15. In a variety store, cards of straight pins cost 16¢ and spools of thread cost 13¢. If Mrs. Barnes spends $1.42 for 2 more spools of thread than cards of pins, how many of each did she buy?

16. *Reduce:* (a) $\dfrac{m^2 - 3m + 2}{2m^2 + m - 3}$ (b) $\dfrac{a^2 - 2a}{a^2 - 4}$

17. *Multiply:* (a) $\dfrac{2x - 4y}{3x^2} \cdot \dfrac{9x^3}{4x^2 - 8xy}$

    (b) $\dfrac{1 - k^2}{2kx} \cdot \dfrac{8k^2}{k^3 - k^4}$

18. Find three consecutive odd integers such that 3 times the middle number plus 3 is equal to the sum of the other two numbers plus 22.

19. *Find S:* $\dfrac{3(x-4)}{4} + \dfrac{2(x+3)}{9} = 5 + \dfrac{2x}{3}$

20. *Name each axiom:*
   (a) $x(y+2) = xy + x(2)$
   (b) $ab + bc = ab + cb$
   (c) $a \cdot 1 = a$
   (d) $x + (y+z) = (x+y) + z$

9 Linear inequalities in two variables*

9-1. HALF-PLANES

When a rectangular coordinate system is set up with an x- and y-axis only, these axes lie in a *plane*. (*Plane* is an undefined term.) If a line is drawn in the plane, it serves to divide the plane into two parts, called *half-planes*. If a point is placed in a plane that contains a line, then that point can be *on the line* or *in the upper half-plane* or *in the lower half-plane*. (If the line is vertical, for example, $x = 2$, it divides the plane into a *left* and *right* half-plane.) (See Figure 9-1.)

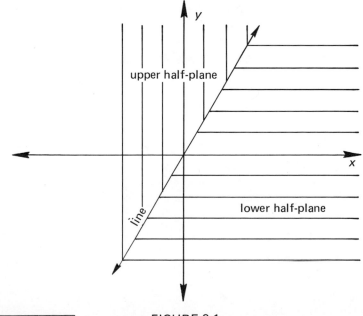

FIGURE 9-1

*Optional chapter.

9-2.   THE GRAPH OF AN INEQUALITY
         IN TWO VARIABLES

In chapter 5 we saw that $A = \{x \mid x > 4\}$ is an infinite set of real
numbers and that the graph was part of a line. Only one variable is
used and the graph is one dimensional.

If an equation or inequality contains a second variable, then the
solution set contains one or more ordered pairs. The set $B = \{(x,y) \mid$
$x + y = 2\}$ is infinite and the graph is a straight line. Every member of
B is on the line $x + y = 2$. Now let us alter B slightly to get set
$B' = \{(x,y) \mid x + y \leq 2\}$. In this case we want all ordered pairs whose
coordinates have a sum *less than or equal to 2*. A little trial and error
work should convince us that many points satisfy the inequality and
that these points are *not* on a straight line. For example, every point
with negative x- and y coordinates is in B'. Also, every point on the
line $x + y = 2$ is in B'. The following theorem is taken from analytic
geometry.

THEOREM 9-1: *The graph of a linear inequality in two variables is a half-plane.
Furthermore, if the inequality is replaced by an equation involving the same
terms, this equation graphs as the boundary line of the half-plane.*

examples: (a)   Graph $S = \{(x,y) \mid x + y \leq 4\}$

This graph is a half-plane with boundary line
$x + y = 4$. First, plot the line $x + y = 4$.

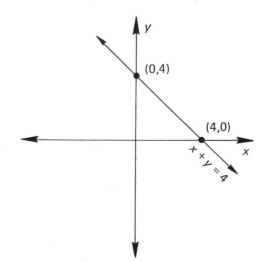

Then, to decide which half-plane (upper or lower) is correct, choose *any* point not on the line and substitute it into $x + y \leq 4$ to see if that point is in the set S. Let us choose $(0,0)$, which is a point below the line: $0 + 0 \leq 4$, which is true, so that $(0,0) \epsilon S$ and every point on the same side (lower) is in S.

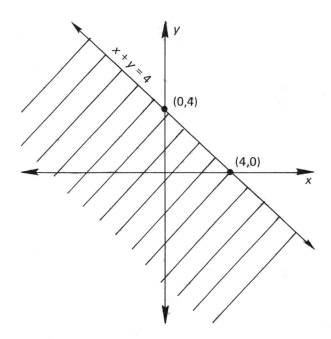

Finally, shade the lower half-plane to indicate that this is the answer.

Note: The boundary line $x + y = 2$ is included in the set, and this is indicated by making the line solid. In the next example the boundary line is not included, and this line will be a dotted line.

(b) Graph $B = \{(x,y) \mid x - 2y > 1\}$

First, plot the boundary line $x - 2y = 1$. (The dotted line indicates that the line is not in the set.)

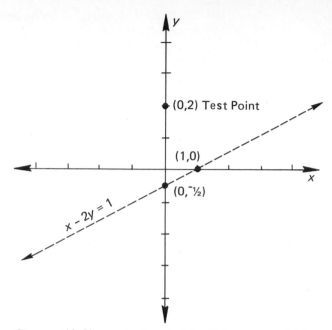

Choose (0,2) as a test point to determine which half-plane is correct:

$$0 - (2)(2) > 1$$
$$^-4 > 1 \text{ is false}$$

which means (0,2) is not in the set and all points on the opposite side of the line are in set B.

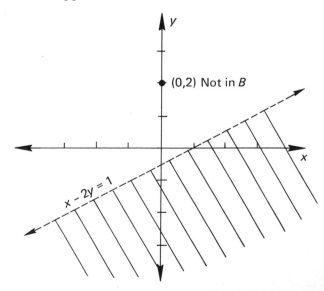

(c)   Graph $C = \{(x,y) \mid x \leq 3\} \cap \{(x,y) \mid y < x + 2\}$

The boundary line for the first set is $x = 3$ (solid line) and all points on the line and in the left half-plane are in this set.

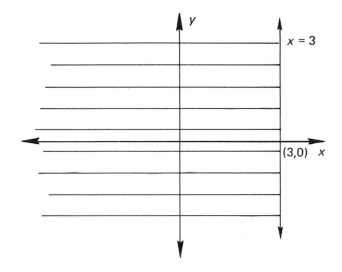

Now plot the second set on the same axes. The boundary line is $y = x + 2$. Use $(0,0)$ as a test point. Since $0 < 0 + 2$ is true, we know that $(0,0)$ is in the second set. Shade the lower half-plane. The intersection of these two sets appears where double shading appears, and the final graph should be shaded in solid.

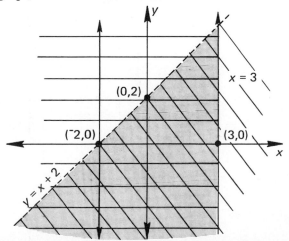

(d) Graph $D = \{(x,y) \mid {}^-2 \leq y \leq 4 \text{ and } 3x + 4y \geq 12\}$

The first portion of the graph lies between the
parallel lines $y = 4$ and $y = {}^-2$.

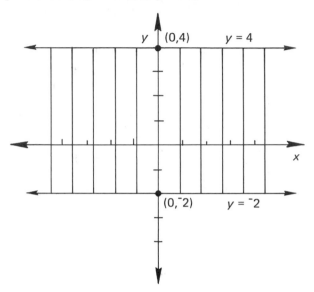

Then plot $3x + 4y = 12$ and choose $(0,0)$ as a
test point.

$$3(0) + 4(0) \geq 12$$
$$0 \geq 12 \text{ is false}$$

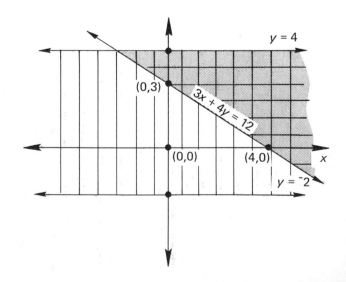

Exercise 60

1. If a point has negative x- and y-coordinates, show the portion of the plane in which it lies.
2. Sketch a half-plane bounded by the line $x = 3$.
3. Sketch another half-plane bounded by line $x = 3$.

Graph each of the following sets on a separate pair of axes:

4. (a) $A = \{x \mid x = 2\}$ (b) $B = \{(x,y) \mid x = 2\}$
   (c) $C = \{(x,y) \mid x \leq y\}$ (d) $D = \{x \mid x \leq 2\}$
5. (a) $E = \{x \mid x + 2 = 6\}$ (b) $F = \{x \mid x + 2 \leq 6\}$
   (c) $G = \{(x,y) \mid x + 2 < 6\}$ (d) $H = \{(x,y) \mid x + 2 > 6\}$
6. $\{(x,y) \mid x + y \geq 7\}$ 7. $\{(x,y) \mid x - y \leq 3\}$
8. $\{x \mid x - 8 = 3\}$ 9. $\{(x,y) \mid y \geq x + 2\}$
10. $\{(x,y) \mid 1 < x < 5\}$ 11. $\{(x,y) \mid 1 < y < 5\}$
12. $\{(x,y) \mid {}^-2 \leq x \leq 6\}$ 13. $\{(x,y) \mid 0 \leq x < 10\}$
14. $\{(x,y) \mid {}^-6 \leq y \leq 0\}$
15. $\{(x,y) \mid 0 \leq y < 3 \text{ and } y \leq x + 5\}$
16. $\{(x,y) \mid {}^-1 \leq x \leq 2 \text{ and } 2 \leq y \leq 5\}$
17. $\{(x,y) \mid x + 4y \geq 8\}$
18. $\{(x,y) \mid x < 2\} \cap \{(x,y) \mid y > 3\}$
19. $\{(x,y) \mid y \leq 3\} \cap \{(x,y) \mid y \geq {}^-3\}$
20. $\{(x,y) \mid x < y + 2\} \cap \{(x,y) \mid x > y - 2\}$
21. $\{(x,y) \mid x + y \geq 4\} \cap \{(x,y) \mid x + y \leq 1\}$
22. $\{(x,y) \mid {}^-2 < x < 4 \text{ and } 1 < y < 3\}$
23. $\{(x,y) \mid y \geq x + 4 \text{ and } x < 5 \text{ and } y < 1\}$
24. $\{(x,y) \mid x + 2y < 6, 0 \leq y \leq 6\}$
25. $\{(x,y) \mid y < 4 \text{ and } y \leq x \text{ and } x \geq {}^-3\}$
26. $\{(x,y) \mid y \geq {}^-3 \text{ and } y > {}^-x + 2 \text{ and } 3y \leq 3x + 1\}$
27. $\{(x,y) \mid 2y \geq x + 4 \text{ and } x < 7 \text{ and } x + y \leq 8\}$
28. Plot the rectangle whose vertices are at $(3,2)$, $(3,{}^-1)$, $({}^-2,2)$, and $({}^-2,{}^-1)$. Then write the set of inequalities which describes the interior region of the rectangle.

Graph the following sets:

29. $\{(x,y) \mid x = 3 \text{ or } y \geq 9\}$
30. $\{(x,y) \mid x < {}^-1 \text{ or } y < 2\}$
31. $\{(x,y) \mid x \geq 0 \text{ and } y \geq 0\} \cup \{(x,y) \mid x \leq 0 \text{ and } y \leq 0\}$
32. $\{(x,y) \mid x - y \geq 2 \text{ or } x + y \geq {}^-2\}$

10 Exponents and radicals

10-1. DEFINITIONS AND THEOREMS ABOUT POSITIVE INTEGRAL EXPONENTS

Recall this definition from section 2-3:

$x^n = x \cdot x \cdot x \cdot \ldots$ (to n factors) if n is a positive integer.

From this definition several theorems can be proved that enable us to perform operations with exponents in a streamlined manner.

In the following theorem, m and n represent positive integers:

THEOREM 10-1: $x^m \cdot x^n = x^{m+n}$
In words: To multiply numbers with the same base, add the exponents.

THEOREM 10-2A: $\dfrac{x^m}{x^n} = x^{m-n}$, *if m is larger than or equal to n.*

THEOREM 10-2B: $\dfrac{x^m}{x^n} = \dfrac{1}{x^{n-m}}$, *if n is larger than m.*
In words: To divide numbers with same base, take the difference between the exponents and place that answer in the part of the fraction that contains the larger exponent.

THEOREM 10-3: $(xy)^m = x^m y^m$
In words: To raise a product to a power, raise each factor to that power.

Note: Theorem 10-3 is very important—learn it well!

THEOREM 10-4: $(x^m)^n = x^{mn}$

In words: To raise a power to a power, multiply the exponents.

THEOREM 10-5: $\left(\dfrac{x}{y}\right)^m = \dfrac{x^m}{y^m}$

In words: To raise a fraction to a power, raise the numerator and denominator to that power.

examples: Simplify each expression.

(a) $x^3 \cdot x^4 = x^{3+4} = x^7$ (Theorem 10-1)

(b) $\dfrac{6x^3 y^2}{3x^4 y^3} = \dfrac{2}{x^{4-3} y^{3-2}} = \dfrac{2}{xy}$ (Theorem 10-2B)

(c) $[x(x+y)]^2 = x^2 (x+y)^2$ (Theorem 10-3)
$= x^2 (x^2 + 2xy + y^2)$
$= x^4 + 2x^3 y + x^2 y^2$

Note: $(x+y)^3$ would not be a problem which follows Theorem 10-3, as $(x+y)^3 = x^3 + 3x^2 y + 3xy^2 + y^3$.

(d) $(x^3)^2 = x^{3 \cdot 2} = x^6$ (Theorem 10-4)

(e) $x^m \cdot x^2 = x^{m+2}$ (Theorem 10-1)

(f) $\dfrac{8^4}{8} = 8^{4-1} = 8^3 = 512$ (Theorem 10-2A)

(g) $\left(\dfrac{4x^3 y^4}{5pq^3}\right)^3 = \dfrac{(4x^3 y^4)^3}{(5pq^3)^3}$ (Theorem 10-5)

$= \dfrac{64x^9 y^{12}}{125p^3 q^9}$ (Theorems 10-3 and 10-4)

Exercise 61

Simplify each expression, unless directed otherwise:

1. $x^2 \cdot x^3$

2. $x^4 (3x^5)$

3. $2a^2 (3ab^3)$

4. $\dfrac{a^3}{a}$

5. $(y^3)^2$

6. $\left(\dfrac{x^2 y^3}{z}\right)^2$

7. $x^k \cdot x^3$

8. $(a^3)^3$

9. $(3a^3)(a^3)$

10. $\dfrac{p^{10}}{p^2}$

11. $\dfrac{p^3}{p^8}$

12. $\dfrac{a^3 x^2 y^3}{axy^5}$

13. $\dfrac{(p^3 q^2)^3}{pq^2}$ 14. $\left(\dfrac{p^3 q^2}{pq^2}\right)^3$ 15. $\dfrac{p^3 q^2}{(pq^2)^3}$

16. $\dfrac{24^3}{20^4}$ 17. $\dfrac{(100)^2}{(50)^3}$ 18. $\dfrac{a^2 - 16}{a - 4}$

19. $\dfrac{3^4 \cdot 3^5}{9^4}$ 20. $\left(\dfrac{21m^3}{7m^2}\right)^2$ 21. $\dfrac{(a + b)^3}{(a + b)^2}$

22. $(x^2 y^2 z^3)^3$ 23. $(5xy^2)^3$ 24. $(a^x)^3$

25. *Find S:* $2(x - 3) = x + 3(4 - x)$

26. *Find S:* $\dfrac{x + 1}{4} + \dfrac{x}{8} = 1$

27. *Find S:*   (a) $(x + 3)(x - 2) = 0$
            (b) $x^2 - 3x + 2 = 0$

28. $\left(\dfrac{2b^3 c^2}{3bc^3}\right)^3$ 29. $[(x + y)^2]^3$ 30. $(3x^2)^k$

31. $\left(\dfrac{a^2 b^2}{c}\right)^3$ 32. $(b^8)^{10}$ 33. $\left(\dfrac{x^2 y^3}{x^2 y}\right)^3$

34. $(5mn)^2$ 35. $(8x^3 y^3 z^2)^2$ 36. $\left(\dfrac{ax}{a^3 x^2}\right)^2$

37. *Find S* $= \{(x,y) \mid (1)\ x - 2y = 3\ \text{and}\ (2)\ x + y = 9\}$.
38. Find the *x*-intercept for the line $3y = x + 8$.
39. *Find S:* $x^2 - 1 = 0$
40. $x^3 (3x^3)(2x^3)$ 41. $(5a^2)(10a^4)$ 42. $(5a^2)^4$
43. $x^k \cdot x^2 \cdot x^4$ 44. $(x^2)^3 x^4$ 45. $(2a)^3 (2a)^2$

46. $\left(\dfrac{1}{3}\right)^2$ 47. $\left(\dfrac{3}{5}\right)^3$ 48. $\left(\dfrac{-3}{2}\right)^3$

49 $\left(\dfrac{1}{x + 3}\right)^2$ 50. $\left(\dfrac{-3}{x + 1}\right)^2$ 51. $\left(\dfrac{x + 4}{(x + 4)^2}\right)^2$

52. $(x^2)^p$ 53. $\left(\dfrac{4x}{y}\right)^3$ 54. $\left(\dfrac{a^x}{b^x}\right)^x$

55. Find two numbers whose sum is 20, and where twice the smaller less the larger equals 7.

56. How many dimes are there in a group of nickels and dimes worth \$2 in which the nickels outnumber the dimes three to one?

10-2.   NEGATIVE AND ZERO INTEGRAL EXPONENTS*

Our discussion of exponents has been limited to exponents that are positive integers. Now we intend to allow negative integers and zero as exponents, and we must determine if they produce numbers that we already have or if they must be treated in a special way.

Our first decision is to insist that negative exponents follow the five theorems of the last section. Now consider x^0 :

$$\text{Observe} \quad 1 = \frac{x}{x} = x^{1-1} = x^0$$

Therefore, $x^0 = 1$, since equality is transitive; and x^0 is not a new number but just another form of the multiplicative identity, 1.

Next consider x^{-1} :

$$(x^{-1})x = x^{-1+1} = x^0 = 1.$$

Here we see that x^{-1} is the number that can be multiplied by x to obtain 1 as a product. This makes x^{-1} the reciprocal of x, as every number has a unique reciprocal. That is,

$$x^{-1} = \frac{1}{x} \, , x \neq 0$$

Likewise, $\quad x^{-3} = \frac{1}{x^3} \quad \text{and} \quad x^{-p} = \frac{1}{x^p}$

Negative exponents are used in scientific work to simplify the writing of very small numbers. (See section 10-3.) However, we shall consider negative exponents poor form and will always simplify an answer that contains a negative exponent.

examples: (a) $3^{-1} = \frac{1}{3}$

(b) $x^{-1} y^{-1} = \frac{1}{x} \cdot \frac{1}{y} = \frac{1}{xy}$

(c) $\dfrac{x^{-3} y^2}{z^{-1}} = \dfrac{\frac{1}{x^3} \cdot y^2}{\frac{1}{z}} = \dfrac{zy^2}{x^3}$

Note: Example (c) indicates that a *factor* may move from numerator to denominator by changing the sign of its exponent.

(d) $(x + y)^{-1} = \dfrac{1}{x + y}$

---
*Optional section.

(e) $\quad xy^{-1} = \dfrac{x}{y}$

(f) $\quad x^{-1} + 2 = \dfrac{1}{x} + 2 = \dfrac{1}{x} + \dfrac{2x}{x} = \dfrac{1 + 2x}{x}$

(g) $\quad \dfrac{x^{-1} + y^{-1}}{x^{-2} + y^{-2}} = \dfrac{\dfrac{1}{x} + \dfrac{1}{y}}{\dfrac{1}{x^2} + \dfrac{1}{y^2}} = \dfrac{xy^2 + x^2 y}{y^2 + x^2}$

Note: The *terms may not move* from numerator to denominator by changing the sign of the exponent.

(h) $\quad 3x^0 = 3 \cdot 1 = 3$

(i) $\quad (3x)^0 = 1$

(j) $\quad (4x^2 y^{-2})^3 = 4^3 x^6 y^{-6} = \dfrac{64x^6}{y^6}$

Exercise 62

Simplify — Reduce all fractions and leave no negative exponents:

1. $(5x)^{-1}$ 2. $3x^{-2}$ 3. ab^{-3}

4. $ax^2 y^{-2}$ 5. $6^{-1} yz^3$ 6. $(xy)^{-2}$

7. $3(a^2 b^2)^{-1}$ 8. $(a + b)^{-2}$ 9. $p^{-3} q^{-4}$

10. $3m^{-k}$ 11. 10^{-6} 12. 14^{-2}

13. $-\dfrac{4x^3 y^2}{(2xy)^3}$ 14. $a^{-1} + b$ 15. $1 + x^{-1}$

16. $(a^2)^{-3}$ 17. $(p^3 q^3)^{-3}$ 18. $a^{-2} \cdot a^{-3}$

19. $x^2 \cdot x^{-3}$ 20. $\dfrac{x^{-4}}{x^2}$ 21. $(3x^{-4})(2x^6)$

22. $\dfrac{3^{-4}}{3^3}$ 23. $\dfrac{10^6}{10^{-2}}$ 24. $\dfrac{10^3}{10^{-1}}$

25. 16^0 26. $12x^0$ 27. $\dfrac{10^4 \cdot 10^5}{10^{-2}}$

28. $(13^5)^0$ 29. 525^{-1} 30. $\dfrac{x^{-2} y^3 z^{-3}}{x^2 y^3 z^3}$

31. $a^{-1} + b^{-1}$

32. $\dfrac{a^{-1} + 2}{a^{-1} - 2}$

33. $2^{-1} + 3^{-2}$

Complete the following statements:

34. To raise a product to a power _____ .

35. To raise a fraction to a power, raise _____ .

36. To _____ raise each factor to that power.

37. The distributive axiom is _____ .

38. The number $\sqrt{2}$ is a(n) _____ number.

39. When the sum and difference of two terms are multiplied, the product is a(n) _____ .

40. The additive identity is _____ .

10-3. SCIENTIFIC NOTATION*

When performing computations with a mixture of large and small numbers, establishing the decimal point in the answer can be a problem. To make this type of work easier, the following system was developed.

The number 365 can be written 3.65×10^2. This latter expression is called the *standard form* or *scientific form* of the number 365. To arrive at scientific form, a number must be written as the product of a number between 1 and 10 and a power of 10. For example,

$$56 = 5.6 \times 10 \text{ and } .0365 = \frac{365}{10,000} = \frac{3.65}{100} = 3.65 \times \frac{1}{10^2} = 3.65 \times 10^{-2}.$$

To put a number in scientific form:

1. Move the decimal point so that only one digit is on its left.
2. Multiply by a power of 10 whose exponent matches the number of places the decimal point was moved. The exponent is positive if the original number is greater than 1, and negative if the number is less than 1.

examples: (a) $2450 = 2.45 \times 10^3$

(b) $48,000,000 = 4.80 \times 10^7$

(c) $.00213 = 2.13 \times 10^{-3}$

---

*Optional section.

A problem like $\dfrac{.0036 \times 560}{.024}$ can be simplified by using scientific notation.

$$\frac{.0036 \times 560}{.024} = \frac{3.6 \times 10^{-3} \times 5.60 \times 10^2}{2.4 \times 10^{-2}}$$

$$= \frac{3.6 \times 5.6 \times 10^{-3} \times 10^2}{2.4 \times 10^{-2}}$$

$$= \frac{3.6 \times 5.6}{2.4} \times 10$$

$$= \frac{3(\cancel{1.2}) \times \cancel{2}(2.8)}{\cancel{2}(\cancel{1.2})} \times 10$$

$$= 8.4 \times 10 = 84$$

Exercise 63

Write each expression without a denominator and simplify:

1. $\dfrac{1}{10^3}$

2. $\dfrac{10}{10^{-4}}$

3. $\dfrac{10^5}{10^{-2}}$

4. $\dfrac{10^{-3}}{10^{-2}}$

5. $\dfrac{1,000}{100,000}$

6. $\dfrac{10^{-2}}{10^{-6}}$

Write in scientific form:

7. 450
8. 6490
9. 12,000
10. .00461
11. 5.26
12. .0984
13. .611
14. 2,000,000
15. 147,000,000
16. $.0024 \times 10^{-3}$
17. 49×10^{-2}
18. 585×10^{-3}
19. 522×10^3
20. 2530×10^{-6}
21. $.999 \times 10^{-1}$

Simplify, using scientific notation:

22. $420 \times .0003$

23. $\dfrac{480 \times 640}{.96}$

24. $\dfrac{63,000 \times 54,000}{2,800 \times 72,000}$

25. $\dfrac{.0009 \times 400,000}{16 \times 8,100}$

26. $\dfrac{1}{70 \times 800 \times 900}$

27. $\dfrac{.0056}{.000036 \times 490}$

Change from scientific form to regular form:

28. 5.34×10^4 29. 6.46×10^{-6} 30. 9.88×10^{-2}

31. 7.98×10^7 32. 1.01×10^{-3} 33. 5.99×10^{10}

10-4.   ROOTS AND RADICALS

To square a number, we simply multiply it by itself. However, if we are to reverse this process, we must take a given number and find one of two equal factors of this number. Such a number is called the *square root* of a given number. We say that 3 is the square root of 9, and 5 is the square root of 25. To find the square root, a new symbol, $\sqrt{}$, called a *radical* is used.

$\sqrt{16} = 4$ reads "the principal square root of 16 equals 4." The 16 inside the radical is called the *radicand*.

Actually, both 4^2 and $(^-4)^2$ equal 16. However, we choose to call $\sqrt{16}$ equal to 4 only.

The operation $\sqrt{^-4}$ cannot be performed for real numbers. There is no real number whose square is negative. If we allow $\sqrt{^-4} = {}^-2$, then $(^-2)^2$ would have to be $(^-4)$. But we know that $(^-2)^2 = 4$. In the remainder of this book, all radicands will be assumed to be non-negative numbers. With this assumption we can write $\sqrt{x^2} = x$. A more precise treatment of this concept will be given in your next algebra course.

Many times a number must be left in radical form, such as $\sqrt{3}$. $\sqrt{3}$ has the property $(\sqrt{3})^2 = 3$, but $\sqrt{3}$ is an irrational number and cannot be expressed without a radical. An approximation of $\sqrt{3}$ can be made to as many decimal places as desired; but $\sqrt{3}$ will never be an exact fraction. By using the table in Appendix B, we see that $\sqrt{3} \approx 1.732$. (The symbol $\approx$ stands for "is approximately equal to.")

Knowing when a radical must be used is a matter of experience gained by becoming familiar with those numbers which are perfect squares.

We shall use radicals only when necessary.

examples: (a) $\sqrt{400} = \sqrt{(20)^2} = 20$

(b) $\sqrt{4x^2} = \sqrt{(2x)^2} = 2x$, if x is non-negative

(c) $\sqrt{81x^4} = \sqrt{(9x^2)^2} = 9x^2$

(d) $\sqrt{16x^8} = \sqrt{(4x^4)^2} = 4x^4$

(e)   Use tables to approximate $7 + \sqrt{10} = 7 + 3.162$
$= 10.162$

Exercise 64

1. If $5^2 = 25$, then $5 = $ _____ .
2. If $a^2 = 49$, then $a = $ _____ .

Simplify: (Assume all variables to be non-negative.)

3. $\sqrt{25}$	4. $\sqrt{100}$	5. $\sqrt{81x^2}$
6. $\sqrt{x^2 y^2}$	7. $\sqrt{25y^4}$	8. $\sqrt{49x^6}$
9. $\sqrt{225}$	10. $\sqrt{1024}$	11. $\sqrt{3600k^8}$

Use Appendix B table to evaluate:

12. $\sqrt{5}$	13. $\sqrt{18}$	14. $^-\sqrt{21}$
15. $\sqrt{68} + 68$	16. $21 - \sqrt{13}$	17. $8(\sqrt{7})$
18. $\dfrac{1 + \sqrt{6}}{2}$	19. $\dfrac{^-3 - \sqrt{3}}{3}$	20. $\dfrac{2 + 3\sqrt{12}}{4}$

10-5. MULTIPLICATION OF RADICALS

THEOREM 10-6: $\sqrt{a} \cdot \sqrt{b} = \sqrt{ab}$

We will make no attempt to give a proof for this theorem. It states that radicals may be multiplied by multiplying their radicands. The importance of this theorem lies in the symmetric form: $\sqrt{ab} = \sqrt{a}\sqrt{b}$. This tells us that if a radicand is factorable, then each factor may be written within a separate radical. The following simplifications are possible.

(a) $\sqrt{18} = \sqrt{9} \cdot \sqrt{2} = 3\sqrt{2}$

(b) $\sqrt{24} = \sqrt{4} \cdot \sqrt{6} = 2\sqrt{6}$

(c) $\sqrt{48} = \sqrt{16} \cdot \sqrt{3} = 4\sqrt{3}$

(d) $\sqrt{x^3} = \sqrt{x^2} \cdot \sqrt{x} = x\sqrt{x}$

(e) $\sqrt{40x^5} = \sqrt{4} \cdot \sqrt{10} \cdot \sqrt{x^4} \cdot \sqrt{x}$
$\qquad = 2 \cdot \sqrt{10} \cdot x^2 \sqrt{x} = 2x^2 \sqrt{10x}$

(f) $6\sqrt{12} = 6\sqrt{4} \cdot \sqrt{3} = 6 \cdot 2\sqrt{3} = 12\sqrt{3}$

(g) $5\sqrt{28} \cdot \sqrt{14} = 5\sqrt{4} \cdot \sqrt{7} \cdot \sqrt{2} \cdot \sqrt{7}$
$\qquad = 5 \cdot 2\sqrt{7} \cdot \sqrt{7} \cdot \sqrt{2}$

$$= 10 \cdot 7\sqrt{2}$$
$$= 70\sqrt{2}$$

Note: Factors outside the radicals may *not* be multiplied by radicands.

Exercise 65

Multiply and simplify:

1. $\sqrt{3} \cdot \sqrt{3}$ 2. $\sqrt{3} \cdot \sqrt{5}$ 3. $\sqrt{7} \cdot \sqrt{5}$

4. $\sqrt{6} \cdot \sqrt{4}$ 5. $\sqrt{8} \cdot \sqrt{6}$ 6. $\sqrt{7} \cdot \sqrt{7}$

7. $3\sqrt{2} \cdot \sqrt{2}$ 8. $2\sqrt{3} \cdot \sqrt{3}$ 9. $5\sqrt{3} \cdot \sqrt{5}$

10. $^-2\sqrt{3} \cdot \sqrt{6}$ 11. $^-3\sqrt{3}\,(2\sqrt{7}\,)$ 12. $11\sqrt{11} \cdot \sqrt{22}$

13. $\sqrt{4x^2}$ 14. $\sqrt{8x^2}$ 15. $\sqrt{x^2 y^2 z}$

16. $(x^3)^3$ 17. $x^3 \cdot x^3$ 18. $\sqrt{x^6}$

19. $(x^2 y^2)^4$ 20. $\left(\dfrac{x^3}{x}\right)^2$ 21. $\dfrac{16x^2 y^3}{8xy^5}$

22. $(\sqrt{5}\,)^2$ 23. $(6\sqrt{3}\,)^2$ 24. $(4\sqrt{6}\,)^2$

25. $1 + 2\sqrt{9}$ 26. $^-4 + 7\sqrt{49}$

27. $\dfrac{8 - 4\sqrt{3}}{4}$ 28. $\dfrac{2x + x\sqrt{3}}{x}$

29. *Find S:* $\quad ^-(x - 2) + 2(1 - 3x) = ^-10$

30. *Find S:* $\quad 3x(x - 2) = 0$

31. *Find S:* $\quad \{(x,y) \mid x - 2y = 1 \quad \text{and} \quad 2x + 2y = 5\}$

32. *Graph* $\quad \{(x,y) \mid x - 2 = 6\}$

33. *Factor:*   (a) $b^2 - b$ (b) $c^3 - c$

              (c) $x^2 - 3x - 18$ (d) $6x^2 + 5x - 4$

              (e) $x^2 + 16$ (f) $2x^3 - 4x^2 + 50x$

Multiply and simplify:

34. $\sqrt{72}$ 35. $\sqrt{200}$ 36. $3\sqrt{125}$

37. $14\sqrt{8}$ 38. $^-3\sqrt{20}$ 39. $^-7\sqrt{25}$

40. $\sqrt{3} \cdot \sqrt{15}$ 41. $\sqrt{6} \cdot \sqrt{18}$ 42. $\sqrt{9} \cdot \sqrt{49}$

43. $90\sqrt{9}$ 44. $9\sqrt{90}$ 45. $(\sqrt{25}\,)^2$

46. $2(\sqrt{2}\,)^2$ 47. $(2\sqrt{2}\,)^2$ 48. $(5\sqrt{3}\,)(\sqrt{6}\,)^2$

49. If $x = {}^-1$, $y = {}^-2$, and $z = 3$, evaluate

    (a) $2y^2$ (b) $(3y)^2$ (c) ${}^-2xy$

    (d) ${}^-x^2$ (e) x^2z (f) xy^2z

    (g) $4xz + 2xy$ (h) $3x^2 + 2y^2 - z$

50. State the commutative axiom for addition.

51. What is the additive inverse of $^-6$?

52. What is found in a solution set?

53. $x^3x^5 + x^2 \cdot x^6 - 3(x^2)^4 = ?$

54. $^-2 + 3 - {}^-6 - 2 + {}^-5 + 7 - 4 + 0 = ?$

55. *Solve each equation for y:* (a) $m = ay - 2$

                                         (b) $y = xy + 1$

Perform the indicated operations:

56. $(x + 4)^2$ 57. $(2x + 3)^2$

58. $(1 + \sqrt{2})^2$ 59. $(3 + \sqrt{5})^2$

60. $(1 + \sqrt{2})(1 - \sqrt{2})$ 61. $(\sqrt{3} - 5)(\sqrt{3} + 5)$

62. *Add:* $\dfrac{1}{x} + \dfrac{1}{x + 1}$ 63. $\sqrt{44x} \cdot \sqrt{24x}$

64. $4\sqrt{25x^3} \cdot 2\sqrt{5x}$ 65. $ab\sqrt{12ab^2} \cdot a\sqrt{8a^2 b^3}$

10-6. DIVISION AND SIMPLIFICATION OF RADICALS

THEOREM 10-7: $\dfrac{\sqrt{a}}{\sqrt{b}} = \sqrt{\dfrac{a}{b}}$

In words: to divide radicals, simply divide the radicands.

This theorem, along with the multiplication theorem, will be proved in your next algebra course.

examples: *Simplify:*

          (a) $\dfrac{\sqrt{6}}{\sqrt{3}} = \sqrt{\dfrac{6}{3}} = \sqrt{2}$

          (b) $\dfrac{1}{\sqrt{3}} = \dfrac{\sqrt{1}}{\sqrt{3}} = \sqrt{\dfrac{1}{3}}$

          (c) $\dfrac{\sqrt{a^3}}{a} = \dfrac{a\sqrt{a}}{a} = \sqrt{a}$

(d) $\dfrac{\sqrt{10}}{5}$ cannot be simplified.

The main type of radical simplification was covered in the last two sections, namely, removing a factor that is a perfect square from the radicand. Expressions like $\sqrt{24}$ we now write as $2\sqrt{6}$.

Another radical considered to be in poor form is one with a denominator in the radicand, e.g. $\sqrt{\tfrac{1}{3}}$. To simplify,

$$\sqrt{\frac{1}{3}} = \sqrt{\frac{1}{3} \cdot \frac{3}{3}} = \sqrt{\frac{3}{9}} = \sqrt{3\left(\frac{1}{9}\right)} = \sqrt{3\left(\frac{1}{3}\right)^2} = \frac{1}{3}\sqrt{3}.$$

Now the radical has no denominator in the radicand. The idea behind this operation is to make the denominator a perfect square, then to remove it from the radical.

Caution: Be certain that the denominator that you remove remains in the denominator outside the radical.

examples: Simplify:

(a) $\sqrt{\dfrac{1}{2}} = \sqrt{\dfrac{1}{2} \cdot \dfrac{2}{2}} = \sqrt{\dfrac{2}{4}} = \dfrac{1}{2}\sqrt{2}$ or $\dfrac{\sqrt{2}}{2}$

(b) $\sqrt{\dfrac{3}{8}} = \sqrt{\dfrac{3}{8} \cdot \dfrac{2}{2}} = \sqrt{\dfrac{6}{16}} = \dfrac{1}{4}\sqrt{6}$ or $\dfrac{\sqrt{6}}{4}$

In example (b), 8 was changed to 16 rather than 64; this saved extra steps.

(c) $\sqrt{\dfrac{3}{x}} = \sqrt{\dfrac{3}{x} \cdot \dfrac{x}{x}} = \sqrt{\dfrac{3x}{x^2}} = \dfrac{1}{x}\sqrt{3x}$ or $\dfrac{\sqrt{3x}}{x}$

(d) $\dfrac{3}{5x}\sqrt{\dfrac{8}{x^3}} = \dfrac{3}{5x} \cdot 2\sqrt{\dfrac{2}{x^3}} = \dfrac{6}{5x}\sqrt{\dfrac{2}{x^3} \cdot \dfrac{x}{x}}$

$= \dfrac{6}{5x}\sqrt{\dfrac{2x}{x^4}} = \dfrac{6}{5x}\left(\dfrac{1}{x^2}\right)\sqrt{2x} = \dfrac{6\sqrt{2x}}{5x^3}$

Finally, we shall not leave radicals in the denominator. A problem like $\dfrac{1}{\sqrt{3}}$ can be changed to $\dfrac{\sqrt{3}}{3}$ by the following steps:

$$\frac{1}{\sqrt{3}} = \frac{1}{\sqrt{3}} \cdot \frac{\sqrt{3}}{\sqrt{3}} = \frac{\sqrt{3}}{\sqrt{9}} = \frac{\sqrt{3}}{3}$$

examples: Simplify:

(a) $\dfrac{2}{\sqrt{5}} = \dfrac{2}{\sqrt{5}} \cdot \dfrac{\sqrt{5}}{\sqrt{5}} = \dfrac{2\sqrt{5}}{\sqrt{25}} = \dfrac{2\sqrt{5}}{5}$

(b) $\dfrac{12a}{\sqrt{a}} = \dfrac{12a}{\sqrt{a}} \cdot \dfrac{\sqrt{a}}{\sqrt{a}} = \dfrac{12a\sqrt{a}}{a} = 12\sqrt{a}$

(c) $\dfrac{2x\sqrt{x}}{\sqrt{90}} = \dfrac{2x\sqrt{x}}{3\sqrt{10}} \cdot \dfrac{\sqrt{10}}{\sqrt{10}} = \dfrac{2x\sqrt{10x}}{3 \cdot 10} = \dfrac{x\sqrt{10x}}{15}$

Summary of Rules for Simplification of Radicals:

1. Leave no *factor* in the radicand which is a perfect square.
2. Leave no denominator in a radicand.
3. Leave no radical in the denominator.

Exercise 66

Perform all indicated operations and simplify:

1. $\dfrac{\sqrt{8}}{\sqrt{4}}$

2. $\dfrac{\sqrt{12}}{\sqrt{3}}$

3. $\dfrac{\sqrt{3x}}{\sqrt{3}}$

4. $\dfrac{3\sqrt{4}}{3}$

5. $\dfrac{2x^2\sqrt{x}}{\sqrt{x}}$

6. $\dfrac{\sqrt{18}}{\sqrt{3}}$

7. $\sqrt{240}$

8. $\sqrt{50x^2}$

9. $\sqrt{300}$

10. $\dfrac{\sqrt{200}}{\sqrt{2}}$

11. $\dfrac{6\sqrt{98}}{\sqrt{14}}$

12. $\dfrac{3a\sqrt{6a^3}}{\sqrt{6a}}$

13. $\dfrac{6\sqrt{12a^2}}{6\sqrt{a}}$

14. $\dfrac{ab\sqrt{8a^2}}{\sqrt{50}}$

15. $\dfrac{^-3m\sqrt{9m^3}}{6\sqrt{m^3}}$

16. $\sqrt{\dfrac{1}{4}}$

17. $\sqrt{\dfrac{16}{5}}$

18. $-\sqrt{\dfrac{11}{12}}$

19. $3\sqrt{\dfrac{1}{3}}$

20. $\dfrac{ax\sqrt{2x}}{\sqrt{x}}$

21. $\sqrt{\dfrac{m}{8}}$

22. $4x^{-1}$

23. $(xy)^{-1}$

24. $2^0(xy^2)^{-1}$

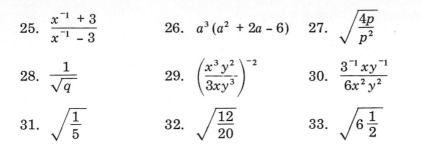

25. $\dfrac{x^{-1} + 3}{x^{-1} - 3}$ 26. $a^3(a^2 + 2a - 6)$ 27. $\sqrt{\dfrac{4p}{p^2}}$

28. $\dfrac{1}{\sqrt{q}}$ 29. $\left(\dfrac{x^3 y^2}{3xy^3}\right)^{-2}$ 30. $\dfrac{3^{-1} xy^{-1}}{6x^2 y^2}$

31. $\sqrt{\dfrac{1}{5}}$ 32. $\sqrt{\dfrac{12}{20}}$ 33. $\sqrt{6\dfrac{1}{2}}$

34. $2(3 + \sqrt{2})$ 35. $\sqrt{3}(1 + \sqrt{3})$ 36. $4\sqrt{2} + \sqrt{6}$

37. *Graph* $\{(x,y) \mid x = 4\}$ 38. *Graph* $\{x \mid x = 4\}$

39. *Graph* $\{(x,y) \mid x + y = 2$ and $3x - y = 6\}$

40. *Find S:*   (1) $4x - 9y = 3$ and   (2) $y = 3$

41. *Find S:*   (1) $2x - y = 6$ and   (2) $y = x + 1$

42. If $x = {}^-9$, $y = 3$,   and $z = 4$, evaluate $3xy^2 - 2xz + z^3$.

43. $\sqrt{\dfrac{2a}{b}}$ 44. $\dfrac{\sqrt{xy}}{\sqrt{y}}$ 45. $3m\sqrt{8m^3}$

46. $\sqrt{48}$ 47. $\dfrac{\sqrt{48}}{8}$ 48. $\dfrac{6 + \sqrt{48}}{8}$

49. $\dfrac{2x + \sqrt{x^3}}{x}$ 50. $a\sqrt{x}(1 + \sqrt{x})$ 51. $ab^2\sqrt{16a^2 b}$

52. To raise a power to a power we must _____ .

10-7. ADDITION OF RADICALS

To add $\sqrt{2}$ and $\sqrt{2}$, two possible answers might come to mind. First, we may try $\sqrt{2 + 2} = \sqrt{4} = 2$. But this proves to be incorrect, as $\sqrt{2} \approx 1.4$, so that $\sqrt{2} + \sqrt{2} \approx 2.8$. Second, we may try calling $\sqrt{2} + \sqrt{2}$ equal to $2\sqrt{2}$. This we can prove by using the distributive axiom:

$$\sqrt{2} + \sqrt{2} = \sqrt{2}(1 + 1) = \sqrt{2}(2) = 2\sqrt{2}$$

Likewise, $8\sqrt{7} + 6\sqrt{7} = \sqrt{7}(8 + 6) = \sqrt{7}(14) = 14\sqrt{7}$

These two examples prompt us to write the following theorem:

THEOREM 10-8: $a\sqrt{b} + c\sqrt{b} = (a + c)\sqrt{b}$

In words: If two radicals have the same radicand, they may be added by adding the coefficients of the radicals.

examples: Simplify and add:

(a) $3\sqrt{5} - 2\sqrt{5} = \sqrt{5}$

(b) $2\sqrt{6} + 3\sqrt{2} + 5\sqrt{6} = 7\sqrt{6} + 3\sqrt{2}$

(c) $\sqrt{50} + \sqrt{18} = 5\sqrt{2} + 3\sqrt{2} = 8\sqrt{2}$

(d) $2x\sqrt{x^3} - 4\sqrt{x^5} = 2x^2\sqrt{x} - 4x^2\sqrt{x} = {}^-2x^2\sqrt{x}$

Exercise 67

Simplify and add if possible:

1. $\sqrt{3} + 2\sqrt{3}$

2. $6\sqrt{5} - 2\sqrt{5} + \sqrt{5}$

3. $a\sqrt{2} + 2a\sqrt{2}$

4. $b\sqrt{x} - 3b\sqrt{x}$

5. $\sqrt{2} + \sqrt{3}$

6. $3\sqrt{3} + \sqrt{3}$

7. $4\sqrt{8} - 3\sqrt{8} - \sqrt{8}$

8. $2xy\sqrt{6} + 3xy\sqrt{6} - 7xy\sqrt{6}$

9. $\sqrt{27} + \sqrt{48}$

10. $4\sqrt{20} - 6\sqrt{125}$

11. $3\sqrt{\dfrac{1}{2}} + 6\sqrt{\dfrac{1}{2}}$

12. $4\sqrt{.03} + 2\sqrt{.03}$

13. $3\sqrt{\dfrac{2}{3}} + 4\sqrt{\dfrac{3}{2}}$

14. $2\sqrt{\dfrac{16}{5}} - 2\sqrt{\dfrac{9}{5}}$

15. $\sqrt{160}$

16. $\sqrt{250} + \sqrt{1000}$

17. $3\sqrt{3}(\sqrt{2} + 3)$

18. $4x(3\sqrt{5} - 4\sqrt{5})$

19. $\sqrt{6}(\sqrt{2} + \sqrt{3})$

20. $a\sqrt{2}(a\sqrt{8} - a\sqrt{9})$

21. $(\sqrt{3} + 1)(\sqrt{3} - 1)$

22. $(2 + \sqrt{5})(2 - \sqrt{5})$

23. *Find S:* $2(x + 3) = 0$

24. *Find S:* $2x(x + 3) = 0$

25. *Find S:* $2x^2 + 5x + 3 = 0$

26. *Add:* $\dfrac{3}{x} + \dfrac{2}{x^2} + 1$

Simplify and add:

27. $3\sqrt{5} + 3\sqrt{45}$

28. $x\sqrt{x} - 2x\sqrt{x}$

29. $\sqrt{5} + \sqrt{\dfrac{1}{5}}$

30. $2\sqrt{6} + \sqrt{\dfrac{1}{6}}$

31. $\dfrac{\sqrt{12}}{5} + \dfrac{\sqrt{75}}{6}$

32. $\dfrac{2\sqrt{45}}{x} + \dfrac{3\sqrt{80}}{2x}$

33. $\dfrac{6 + 3\sqrt{16}}{8}$ 34. $\dfrac{2 - 3\sqrt{12}}{2}$

35. $\dfrac{1 - x^2}{1 + x}$ 36. $\dfrac{x - x^2}{1 - x^2}$

37. How did we define subtraction, $a - b$?

38. Find the multiplicative inverse (reciprocal) of $\dfrac{3}{5}$.

39. Distinguish a theorem from an axiom.

40. What is the role of definitions in a mathematical system?

41. Is $(3,0) \in \{(x,y) \mid y = 0\}$?

42. Factor $a^2 + 16$.

10-8. FURTHER SIMPLIFICATION OF RADICALS

One of the first axioms taken in this course states that the real numbers are closed under the operations of addition and multiplication. This means that every sum or product of two real numbers is itself a real number. The closure property does not hold for all subsets of the real numbers. Take, for example, the set of all irrational numbers. If $\sqrt{2}$ is multiplied by $\sqrt{2}$, the product is 2, which is not an irrational number. Therefore, the set of irrational numbers is not closed with respect to multiplication. It was this discovery in section 10-6 which allowed us to consider an expression like $\dfrac{1}{\sqrt{3}}$ to be in poor form, and to change it to $\dfrac{\sqrt{3}}{3}$.

Now we turn our attention to an expression like $\dfrac{1}{2 + \sqrt{3}}$. This also is in poor form because it contains a radical in the denominator. What we must do is find a number to multiply by $2 + \sqrt{3}$ which will give a product that is a rational number (an integer if possible). Three candidates for such a factor are suggested. They are $\sqrt{3}$, $2 + \sqrt{3}$, and $2 - \sqrt{3}$. Trying each one in turn, we find:

(a) $(2 + \sqrt{3})\sqrt{3} = 2\sqrt{3} + 3$ still irrational

(b) $(2 + \sqrt{3})(2 + \sqrt{3}) = 4 + 4\sqrt{3} + 3 = 7 + 4\sqrt{3}$ irrational

(c) $(2 + \sqrt{3})(2 - \sqrt{3}) = 4 - 3 = 1$ success

Part (c) shows us two irrational numbers $2 + \sqrt{3}$ and $2 - \sqrt{3}$ which are called *conjugates* of one another.

DEFINITION: *$\sqrt{x} - \sqrt{y}$ is the conjugate of $\sqrt{x} + \sqrt{y}$;*
                 also, $x - \sqrt{y}$ is the conjugate of $x + \sqrt{y}$.

Returning to the simplification of $\dfrac{1}{2+\sqrt{3}}$, we can multiply by

$\dfrac{2-\sqrt{3}}{2-\sqrt{3}}$ and obtain $\dfrac{1}{(2+\sqrt{3})} \cdot \dfrac{(2-\sqrt{3})}{(2-\sqrt{3})} = \dfrac{2-\sqrt{3}}{4-3} = \dfrac{2-\sqrt{3}}{1} = 2-\sqrt{3}$

examples: Simplify:

(a) $\dfrac{\sqrt{2}}{1-\sqrt{2}} = \dfrac{\sqrt{2}}{(1-\sqrt{2})} \cdot \dfrac{(1+\sqrt{2})}{(1+\sqrt{2})} = \dfrac{\sqrt{2}+2}{1-2} = \dfrac{\sqrt{2}+2}{-1} = -\sqrt{2}-2$

(b) $\dfrac{\sqrt{5}-\sqrt{3}}{\sqrt{5}+\sqrt{3}} = \dfrac{(\sqrt{5}-\sqrt{3})}{(\sqrt{5}+\sqrt{3})} \cdot \dfrac{(\sqrt{5}-\sqrt{3})}{(\sqrt{5}-\sqrt{3})}$

$\qquad = \dfrac{5-2\sqrt{15}+3}{5-3} = \dfrac{8-2\sqrt{15}}{2} = 4-\sqrt{15}$

Exercise 68

Simplify:

1. $\dfrac{3}{\sqrt{6}}$

2. $\dfrac{2}{\sqrt{8}}$

3. $\dfrac{ax}{\sqrt{x}}$

4. $\dfrac{1}{1+\sqrt{2}}$

5. $\dfrac{9}{3+\sqrt{6}}$

6. $\dfrac{1}{\sqrt{x}-1}$

7. $\dfrac{\sqrt{2}}{\sqrt{2}-1}$

8. $\dfrac{\sqrt{7}}{1-\sqrt{7}}$

9. $\dfrac{a+\sqrt{b}}{\sqrt{b}}$

10. $\sqrt{\dfrac{x+1}{1}}$

11. $\sqrt{\dfrac{3}{1-x}}$

12. $\dfrac{\sqrt{y}}{\sqrt{y}+3}$

13. $\sqrt{150}$

14. $\sqrt{280}$

15. $\sqrt{1000}$

16. $\dfrac{\sqrt{3}+\sqrt{8}}{\sqrt{2}+3}$

17. $\dfrac{a-\sqrt{x}}{a+\sqrt{x}}$

18. $\dfrac{2+3\sqrt{5}}{1-\sqrt{5}}$

19. $\sqrt{x+\dfrac{1}{x}}$

20. $\sqrt{m+\dfrac{1}{m^2}}$

21. $\sqrt{4a^2+\dfrac{1}{a}}$

22. $(x^2 y^2 z^3)^2$

23. $\dfrac{x^3 y}{x^{-3} y^2}$

24. $(ab^x)^3$

25. $(a + x)^2$ 　　　　　26. $(a^2 + x^2)^2$ 　　　　27. $(a + \sqrt{x})^2$

28. $\sqrt{\dfrac{x}{x + 1}}$ 　　　　29. $\sqrt{\dfrac{x}{\sqrt{x} + 1}}$ 　　　　30. $\sqrt{\dfrac{\sqrt{3}}{\sqrt{2} + 3}}$

31. $3\sqrt{3} + 4\sqrt{3}$ 　　　　　　　　32. $x\sqrt{y} - 3x\sqrt{y}$

33. $\sqrt{20} - 2\sqrt{45} + \sqrt{48}$ 　　　　34. $9\sqrt{9} + 6\sqrt{4}$

35. $(a + \sqrt{2})(a - \sqrt{2})$ 　　　　36. $(a + \sqrt{2})(a + \sqrt{2})$

37. $\dfrac{1 + \dfrac{1}{x}}{1 - \dfrac{1}{x}}$ 　　　　　　　38. $\dfrac{3}{1 - \dfrac{1}{9}}$

39. *Find S:* $x^2 - 4x - 5 = 0$

40. *Find S:* $\{(x,y) \mid (1)\ 2x - y = 3 \ \text{ and } \ (2)\ x = y - 3\}$

Cumulative Review: Chapters 1 through 10

1. *Graph* $\{x \mid x = 3 \ \text{ or } \ x = {}^-2 \ \text{ or } \ x = \sqrt{15}\,\}$
2. *Graph* $\{(x,y) \mid x = 3 \ \text{ and } y = 2\}$
3. Name three undefined terms used in this book.
4. State three axioms and a numerical example of each.
5. (a) How many terms are in this radicand? $\sqrt{x^2(x + y)}$
   (b) How many factors?
   (c) Can this radical be simplified?
6. Give an example of a fourth-degree polynomial with two variables.
7. *Square* $(3x - 8)$
8. *Add* $\dfrac{3}{5}$ and $\dfrac{2}{x}$
9. *Reduce:* (a) $\dfrac{2x^2 + x - 1}{2x^2 - 2}$ 　　　　(b) $\dfrac{km - m}{k^2 - k}$
10. How many members are in each set?
    (a) $\{x \mid x = 3 - 7\}$
    (b) $\{x \mid x = 3 \ \text{ and } \ x = {}^-6\}$
    (c) $\{(x,y) \mid y = x + 2\}$
    (d) $\{(x,y) \mid x = 4\}$
    (e) $\{(x,y) \mid x = 4 \ \text{ and } y = {}^-8\}$

11. *Simplify:*   (a) $\sqrt{2 + \dfrac{1}{2}}$ (b) $\dfrac{\frac{1}{2} + 3}{x + \frac{1}{2}}$

   (c) $(a^2 b^3 c)^3$ (d) $^-3p^2 q + 2pq^2 - 6p^2 q$

12. *Solve for b:* (a) $2M = by + 1$ (b) $k = b + bx$

   (c) $\dfrac{1}{b} = a + \dfrac{1}{x}$

13. *Factor completely:* (a) $ax^2 + 4a$ (b) $ax^2 - 4a$
   (c) $ax^2 + 4ax + 4a$
   (d) $ax^3 + 3ax - 70a$

14. If 320 bricks are needed for 24 lineal feet of a wall, how many bricks are needed for 132 lineal feet? (Use proportion.)

15. A 63-inch length of ribbon is cut into two pieces so that one piece is 2½ times the other. How long is each piece? (Use two variables.)

16. Dean has 20 more dimes than quarters, and 3 less nickels than dimes. If his money is worth $5.25, how many of each type of coin does he have?

17. A 40-pound weight and a 90-pound weight are attached to the opposite ends of a 26-foot balance board. Where must the fulcrum be placed in order to balance the two weights?

18. Give an example of a difference of two squares with a common factor.

19. *Find S:* $b^2 - 3b + 2 = 0$

20. *Find S:*   (1) $3x + 4y = 18$ (2) $4x - 2y = 2$

21. *Find S:* $\dfrac{3(x - 6)}{4} - \dfrac{2(x + 1)}{5} = \dfrac{41}{10}$

22. *Simplify:*   (a) $\sqrt{440}$ (b) $\sqrt{28a^2 b^3}$

   (c) $\dfrac{3}{\sqrt{5}}$ (d) $\sqrt{\dfrac{3}{5}}$

23. *Simplify:*   (a) $\dfrac{1}{\sqrt{6} - 1}$ (b) $\dfrac{\sqrt{3}}{\sqrt{3} + 7}$

   (c) $\sqrt{\dfrac{1}{\sqrt{2}}}$ (d) $\left(\dfrac{x^5 y^2}{x^3 y^3}\right)^3$

24. Find two fractions whose product is an integer.

25. What special name is given zero?

11

Quadratic equations

11-1. DEFINITIONS

In this chapter we shall study quadratic (second-degree) equations in one variable. Since there is only one variable, the solutions are single numbers and not ordered pairs. Each equation has two roots (not always distinct).

The *general form of a quadratic equation* is $ax^2 + bx + c = 0$, where a, b, and c are real numbers and $a \neq 0$. The solution set is $S = \{x \mid ax^2 + bx + c = 0\}$. If $b = 0$, then the equation $ax^2 + c = 0$ is called a *pure or incomplete quadratic equation.*

examples:
    (a) $3x^2 + 6x + 7 = 0$ is a complete quadratic equation in general form.

    (b) $2x - 1 = 5x^2$ is not in general form and should be rewritten $5x^2 - 2x + 1 = 0$.

    (c) $2x^2 - 3 = 0$ is a pure quadratic equation.

11-2. AN ALGORITHM FOR EXTRACTING SQUARE ROOTS

Many times the roots of a quadratic equation will contain irrational numbers involving square roots. An approximation of that number may be necessary and sometimes tables are not available or do not contain enough decimal places. The following algorithm will show how to compute the approximate square root of any real number to as many decimal places as desired.

To find $\sqrt{1849}$, begin by finding the decimal point; moving to the left of the decimal point, divide the radicand into blocks of two digits each.

Find the largest number whose square is less than or equal to 18. Place a 4 over the first block of numbers.

$$\sqrt{\overline{18}'\overline{49}}\;\overset{4}{}$$

Square 4 and write 16 below the
18.

$$\begin{array}{r} 4 \\ \sqrt{18'49.} \\ 16 \end{array}$$

Subtract 16 from 18 and bring
down the next *two* digits.

$$\begin{array}{r} 4 \\ \sqrt{18'49.} \\ 16 \\ \hline 249 \end{array}$$

Double the 4 and treat the 8 as
an 80. Place the 8 ☐ next to the
249.

$$\begin{array}{r} 4 \\ \sqrt{18'49.} \\ 16 \\ \hline 8\;\square \;|\; 249 \end{array}$$

Using the 8 ☐ as a trial divisor,
divide 249 by 80. This gives about
3 as a quotient. Place the 3 in the
box to make the divisor 83. Also,
place the 3 over the next block of
digits that was just brought down,
49.

$$\begin{array}{r} 4\;\;3 \\ \sqrt{18'49.} \\ 16 \\ \hline 8\;\boxed{3}\;|\; 249 \end{array}$$

Multiply 3 by 83 and write 249
below 249 and subtract. Since the
remainder is zero, 43 is the exact
square root of 1849. The decimal
point is moved into the root.

$$\begin{array}{r} 4\;\;3 \\ \sqrt{18'49.} \\ 16 \\ \hline 8\;\;3\;|\; 249 \\ 249 \\ \hline 0 \end{array}$$

Check: 43 · 43 = 1849

examples: (a)   Find $\sqrt{1.849}$ to three decimal places.

Since each two decimal places in the radicand
gives only one place in the root, it is necessary
to annex three zeros.

$$\begin{array}{r} 1.\;\;3\;\;5\;\;9 \\ \sqrt{1.'84'90'00} \\ ^{-}1 \\ \hline 2\;\boxed{3}\;|\;0\;84 \\ 69 \\ \hline 26\;\boxed{5}\;\;|1590 \\ 1325 \\ \hline 270\;\boxed{9}\;\;|\;26500 \\ 24381 \\ \hline 2119 \end{array}$$

This problem will probably never come out even and we have $\sqrt{1.849} \approx 1.359$.

Check: $(1.359)(1.359) = 1.846881 \approx 1.847$

(b) Find $\sqrt{.0226}$ to four decimal places.

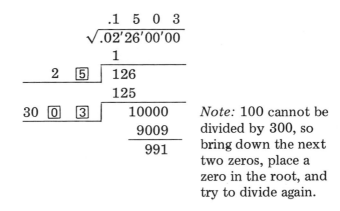

$$.1\ 5\ 0\ 3$$
$$\sqrt{.02'26'00'00}$$

Note: 100 cannot be divided by 300, so bring down the next two zeros, place a zero in the root, and try to divide again.

Check: $(.1503)(.1503) = .02259009 \approx .0226$

Exercise 69

Give the degree and number of variables for each equation defining the set:

1. $\{x \mid 2x - 3 = 7\}$
2. $\{x \mid x^2 + 3x + 7 = 0\}$
3. $\{(x,y) \mid x^2 + y^2 = 8\}$
4. $\{(x,y) \mid x + y = 3\}$
5. $\{(x,y) \mid y = 2\}$
6. $\{(x,y) \mid x^2 = 8\}$

Use the method shown in this section to compute each number to three decimal places, if possible:

7. $\sqrt{169}$
8. $\sqrt{2025}$
9. $\sqrt{6084}$

10. $\sqrt{12.96}$
11. $\sqrt{.3025}$
12. $\sqrt{.0676}$

13. $\sqrt{5.61}$
14. $\sqrt{6124}$
15. $\sqrt{.000314}$

16. $\sqrt{9.0031}$
17. $\sqrt{25.33}$
18. $\sqrt{723.4}$

19. $\sqrt{7.234}$
20. $\sqrt{72.34}$
21. $\sqrt{7234}$

22. Study the answers from problems 18, 19, 20, and 21, and see if you can give the following answers by merely placing the decimal point in the proper place:

    (a) $\sqrt{723400}$ (b) $\sqrt{72340}$

    (c) $\sqrt{.00007234}$ (d) $\sqrt{.0007234}$

23. Find $S = \{(x,y) \mid 2x - y = 18 \quad \text{and} \quad 3x + y = 7\}$

24. Find $S = \left\{x \mid \dfrac{3}{5}x = \dfrac{1}{6}\right\}$

25. Find $S = \{x \mid (x + 3)(x - 7) = 0\}$

11-3. PURE QUADRATIC EQUATIONS OF FORM $ax^2 + c = 0$

An equation like $x^2 = 4$, a quadratic equation in one variable, has two roots. These two roots are the numbers whose squares are 4. They are 2 or $^-2$, since $(2)^2 = 4$ and $(^-2)^2 = 4$. We shall agree that the numbers 2 or $^-2$ can be written ± 2.

To find members $x \in \{x \mid 3x^2 = 8\}$,

$$3x^2 = 8$$

$$x^2 = \frac{8}{3} \qquad \text{(Roots are numbers whose squares are } 8/3 \text{.)}$$

$$x = \pm \sqrt{\frac{8}{3}}$$

$$x = \pm 2\sqrt{\frac{2}{3}} = \pm \frac{2}{3}\sqrt{6}$$

$$S = \left\{\frac{2}{3}\sqrt{6}, -\frac{2}{3}\sqrt{6}\right\} \text{ or } \left\{\pm \frac{2}{3}\sqrt{6}\right\}$$

example: Find $S = \{x \mid 15x^2 - 80 = 0\}$

$$15x^2 - 80 = 0$$

$$3x^2 - 16 = 0$$

$$3x^2 = 16$$

$$x^2 = \frac{16}{3}$$

$$x = \pm \sqrt{\frac{16}{3}}$$

$$x = \pm 4\sqrt{\frac{1}{3}} = \pm \frac{4\sqrt{3}}{3}$$

$$S = \left\{ \pm \frac{4\sqrt{3}}{3} \right\}$$

Check:

If $x = \dfrac{4\sqrt{3}}{3}$

$$15x^2 - 80 = 0$$

$$15\left(\frac{4\sqrt{3}}{3}\right)^2 - 80 = 0$$

$$15\left(\frac{16 \cdot \cancel{3}}{3 \cdot \cancel{3}}\right) - 80 = 0$$

$$5 \cdot 16 - 80 = 0$$

$$80 - 80 = 0$$

$$0 = 0$$

If $x = \dfrac{{}^-4\sqrt{3}}{3}$

$$15x^2 - 80 = 0$$

$$15\left(\frac{{}^-4\sqrt{3}}{3}\right)^2 - 80 = 0$$

$$15\left(\frac{16 \cdot 3}{3 \cdot \cancel{3}}\right) - 80 = 0$$

$$5(16) - 80 = 0$$

$$80 - 80 = 0$$

$$0 = 0$$

Exercise 70

Find S: (Leave all irrational roots in radical form.)

1. $x^2 = 9$
2. $4x^2 = 32$
3. $5x^2 = 10$
4. $2x^2 = 36$
5. $6x^2 = 216$
6. $5x^2 - 20 = 0$
7. $11a^2 - 3 = 0$
8. $2m^2 - 2 = 0$
9. $121b^2 - 144 = 0$
10. $14x^2 - 49 = 0$
11. $\dfrac{2}{3}x^2 = 10$
12. $\dfrac{3}{7}m^2 = \dfrac{1}{2}$

Find the two members of each set:

13. $\{x \mid ax^2 = 4\}$
14. $\left\{x \mid x^2 = \dfrac{1}{4}\right\}$
15. $\{a \mid 2a^2 = 200\}$
16. $\{q \mid 4q^2 = 2q^2 + 6\}$
17. $\{b \mid ab^2 - 5 = 2\}$
18. $\{k \mid 6k^2 + 7 = 3k^2 + 9\}$

Simplify:

19. $(18)^{-1}(9)$
20. $\sqrt{192x^3}$

21. $\dfrac{k + \dfrac{1}{k}}{1 - k}$

22. $\dfrac{4m^2 + 16m}{m^2 + 9m + 20}$

Find S:

23. $2x^2 + 5x^2 - 3 = 11$

24. $\dfrac{3}{8}x^2 = 1$

25. $x + 2 - 3(x + 2) = 0$

26. $3x(x + 2) = 0$

27. $\dfrac{x^2}{4} = 9$

28. $\dfrac{3x^2}{9} = 21$

29. What is:

    (a) $\sqrt{720}$ to two decimal places?

    (b) the square root of $(x + 3)^2$?

    (c) the square of $(x + 3)$?

    (d) the difference when $8x^2 + 3x - 4$ is subtracted from $4x^2 - 5x + 3$?

    (e) the additive inverse of ⁻6?

    (f) the associative axiom for addition?

    (g) the definition of addition?

    (h) the definition of an axiom?

30. *Find S:* $20 = x^2$

11-4. SOLUTION BY FACTORING—A SECOND LOOK

On many occasions in this book we have solved quadratic equations that contained factorable polynomials. In section 6-12 a theorem was given to justify solution by factoring. The principal idea is to get a product equal to zero and then set each factor equal to zero. An equation of degree higher than two can be solved if the factors are given or can be obtained.

examples: (a)   Find $S = \{x \mid x^3 + 2x^2 - 3x = 0\}$

$$x^3 + 2x^2 - 3x = 0$$
$$x(x^2 + 2x - 3) = 0$$
$$x(x + 3)(x - 1) = 0 \quad \text{which means}$$
$$x = 0 \;\text{ or }\; x + 3 = 0 \quad \text{ or }\quad x - 1 = 0$$
$$x = {}^-3 \qquad\qquad x = 1$$

Therefore, $S = \{0, {}^-3, 1\}$

(b)   Find $S = \{x \mid (x^2 - 1)(x^2 + 2x)(x^2 + 5x + 4) = 0\}$

$$(x^2 - 1)(x^2 + 2x)(x^2 + 5x + 4) = 0$$
$$(x + 1)(x - 1)(x)(x + 2)(x + 4)(x + 1) = 0$$

which means

$x + 1 = 0$	or	$x - 1 = 0$	or	$x = 0$	or	$x + 2 = 0$
$x = {}^-1$		$x = 1$		$x = 0$		$x = {}^-2$

or	$x + 4 = 0$	or	$x + 1 = 0$
	$x = {}^-4$		$x = {}^-1$

(Here we find six roots with two of them equal to ⁻1. When two roots are equal, the number is called a *double root*.)

Therefore, $S = \{{}^-1 \text{ (double)}, 1, 0, {}^-2, {}^-4\}$

Now let us reverse the situation. Suppose we know the solution set S and wish to find an equation to which it belongs.

Let $S = \{{}^-1, 3\}$. This means

$x = {}^-1$	or	$x = 3$	which in turn tells us
$x + 1 = 0$		$x - 3 = 0$	

$(x + 1)(x - 3) = 0 \cdot 0$ by the multiplication of
equals axiom

$x^2 - 2x - 3 = 0$ is the equation being
sought.

Exercise 71

Find S:

1. $(x + 5)(x - 8) = 0$
2. $(2x + 1)(x + 3) = 0$
3. $(2x - 3)(8x + 1) = 0$
4. $4x(x - 1) = 0$
5. $x(x + 1)(x + 2) = 0$
6. ${}^-x(x - 7)(2x + 3) = 0$
7. $x^2 - 9x + 8 = 0$
8. $x^2 - 6x + 9 = 0$
9. $4x^2 = 9$
10. $mx^2 - 5mx - 50m = 0$
11. $x^2 + 4x - 10 = 2$
12. $x^2 - 6x = 0$
13. $x^2 - 6x = 7$
14. $x^2 + 8x - 33 = {}^-13$
15. $16y^2 - 2y = 3$
16. $4x^2 + 16x - 9 = 0$
17. $3x^2 - 9 = 0$
18. $a^2 = 21$
19. $x(x + 7) = {}^-6$
20. $3x(x + 2) = {}^-3$
21. $ax^2 - 4 = 0$
22. $2m^2 - m - 3 = 0$

23. $\dfrac{x^2}{2} + \dfrac{11x}{2} + 5 = 0$ 24. $\dfrac{x^2}{7} - \dfrac{11x}{7} = {}^-4$

25. (1) $x - y = 3$ 26. $3x^2 + x - 22 = 2x^2 - 4x - 8$
    (2) $x + 2y = 11$

Find an equation, with x as the variable, which has the given solution set:

27. $S = \{1, 2\}$ 28. $S = \{3, {}^-5\}$
29. $S = \{{}^-4, {}^-2\}$ 30. $S = \{0, 3\}$
31. $S = \{{}^-2, 2, 3\}$ 32. $S = \{5, {}^-5, 0\}$

33. $S = \left\{\dfrac{1}{4}, \dfrac{1}{2}\right\}$ 34. $S = \left\{\dfrac{{}^-3}{5}, \dfrac{1}{8}\right\}$

35. $S = \left\{\dfrac{\sqrt{2}}{3}, {}^-\dfrac{\sqrt{2}}{3}\right\}$ 36. $S = \{1 + \sqrt{3}, 1 - \sqrt{3}\}$

Simplify:

37. $(x^3)^4$ 38. $(a^2 + 3)^2$ 39. $ax(x + 2)^2$

40. Write the general quadratic equation in one variable.
41. Find S if $(x^2 - 9)(x^2 - 3x)(x^2 + 7x + 6) = 0$
42. Find S if $x^3(x^2 + 5x - 66)(x^3 - 25x) = 0$

11-5.   SOLUTION BY COMPLETING THE SQUARE

We have seen that the solution of
$$x^2 = 16 \text{ is } x = \pm 4$$
A similar equation $(x - 3)^2 = 16$ has solution
$$(x - 3) = \pm 4 \quad \text{which is two equations}$$
$$x - 3 = 4 \quad \text{or} \quad x - 3 = {}^-4$$
$$x = 7 \qquad\qquad x = {}^-1$$

This solution proves to be rather simple. The problem is to get the binomial square $(x - 3)^2$. When a binomial is squared, such as $(x - 3)^2 = x^2 - 6x + 9$, the resulting trinomial begins and ends with a square and its middle term is a result of doubling the product of the square roots of the first and last term.

Let us attempt to make up a trinomial square from the binomial $x^2 + 8x$. We are seeking a third term which is the square of one-half the coefficient of x, that is $(4)^2$. We claim that $x^2 + 8x + 16$ is a perfect square.

examples: (a) Complete the square for $x^2 - 2x$. To do this, take ½ of ⁻2 and square it to obtain 1. The completed square is $x^2 - 2x + 1 = (x - 1)^2$.

(b) Complete the square for $x^2 + 3x$.

Take ½ of $3 = \frac{3}{2}$; square this to get ¾.

The completed square is $x^2 + 3x + \frac{9}{4} = (x + \frac{3}{2})^2$.

In the above examples, we changed the given polynomial by adding a third term to it. However, if the polynomial appears in an equation, the same quantity must be added to each side of the equation. The following examples illustrate the method of solving a quadratic equation by completing the square.

examples: (a) Find S if $x^2 - 6x + 5 = 0$

$$x^2 - 6x + 5 = 0$$
$$x^2 - 6x \quad = {}^-5 \quad \text{Now complete the square in the left-hand member}$$
$$x^2 - 6x + 9 = {}^-5 + 9 \quad \text{Add 9 to each side}$$
$$(x - 3)^2 = 4$$
$$x - 3 = \pm 2$$
$$x = {}^-3 \pm 2 \quad \text{which means}$$

$$\begin{array}{ccc} x = {}^-3 + 2 & & x = {}^-3 - 2 \\ x = {}^-1 & \text{or} & x = {}^-5 \end{array}$$

$$S = \{{}^-1, {}^-5\}$$

(b) Find $S = \{x \mid x^2 + 5x + 2 = 0\}$

$$x^2 + 5x + 2 = 0$$
$$x^2 + 5x \quad = {}^-2$$
$$x^2 + 5x + \left(\frac{5}{2}\right)^2 = {}^-2 + \left(\frac{5}{2}\right)^2 = \frac{{}^-8}{4} + \frac{25}{4}$$
$$\left(x + \frac{5}{2}\right)^2 = \frac{17}{4}$$
$$x + \frac{5}{2} = \pm\sqrt{\frac{17}{4}} = \frac{\pm\sqrt{17}}{2}$$
$$x = \frac{{}^-5}{2} + \frac{\pm\sqrt{17}}{2}$$
$$x = \frac{{}^-5 \pm \sqrt{17}}{2}$$

$$S = \left\{ \frac{-5 \pm \sqrt{17}}{2} \right\}$$

(c)   Find $x \in S = \{x \mid 3x^2 + 8x + 1 = 0\}$

$3x^2 + 8x + 1 = 0$

$3x^2 + 8x \quad = {}^-1$

The coefficient of x^2 must be 1, otherwise it is very difficult to determine what number to add to complete the square.

$$x^2 + \frac{8}{3}x \qquad = \frac{{}^-1}{3} \quad \text{Divide by 3}$$

$$x^2 + \frac{8}{3}x + \left(\frac{4}{3}\right)^2 = \frac{{}^-1}{3} + \left(\frac{4}{3}\right)^2$$

$$\left(x + \frac{4}{3}\right)^2 = \frac{{}^-3}{9} + \frac{16}{9} = \frac{13}{9}$$

$$x + \frac{4}{3} = \pm \sqrt{\frac{13}{9}} = \frac{\pm \sqrt{13}}{3}$$

$$x = \frac{{}^-4 \pm \sqrt{13}}{3}$$

$$S = \left\{ \frac{{}^-4 \pm \sqrt{13}}{3} \right\}$$

Exercise 72

Find S by completing the square:

1. $x^2 + 2x - 3 = 0$ 2. $x^2 - 4x - 5 = 0$
3. $x^2 - 8x - 20 = 0$ 4. $x^2 + 10x + 21 = 0$
5. $x^2 + 5x + 6 = 0$ 6. $x^2 - 3x + 2 = 0$
7. $x^2 - 7x - 5 = 0$ 8. $x^2 + 12x - 9 = 0$
9. $2x^2 - 6x + 3 = 0$ 10. $5x^2 + 3x - 4 = 0$

11. $m^2 - \frac{3}{5}m = \frac{1}{2}$ 12. $k^2 + \frac{2}{3}k - 1 = 0$

Find S by any method:

13. $2x^2 - 18 = 0$ 14. $x^2 + 4x + 3 = 0$

15. $2x(x - 4)(x^2 - 25) = 0$ 16. $\frac{1}{2}x^2 = \frac{1}{8}$

17. $\frac{x}{2} + \frac{3x}{4} = 5$ 18. (1) $x - y = 3$
    (2) $2x - 3y = 16$

19. $2x^2 - 5x = 0$ 20. $2x^2 - 5 = 0$

21. *Reduce:*   (a) $\dfrac{xy - y}{ax - a}$ (b) $\dfrac{2x^2(x^2 - 4)}{x^2 - 2x}$

22. *Add:*   (a) $\dfrac{x}{3} + \dfrac{x}{4} + \dfrac{x}{6}$ (b) $\dfrac{x + 4}{x^2 - 1} - \dfrac{x + 2}{x + 1}$

23. *Simplify:* (a) $\dfrac{\frac{1}{3}}{\frac{3}{5}}$ (b) $\sqrt{40xy^2}$

24. *Find S:* $6x^2 + 7x - 20 = 0$

25. What is $\{3, 6, 9, 12\} \cap \{2, 4, 6, 8, 10, 12, 14\}$?

26. If $x = 3 + \sqrt{2}$, evaluate
    (a) $x^2 + 3x + 1$ (b) $2x^2 - x - 5$
    (c) $3x^2 + 4(x + 2) - 3$
    (d) $x^2 + \frac{1}{2}x - 3$

27. *Find S:*   (a) $x^2 = 1$ (b) $x^2 = 9$ (c) $x^2 = \dfrac{9}{16}$

28. *Graph* $\{(x,y) \mid 3x + 2y = 12\}$ by finding the x- and y-intercepts.

11-6.   THE QUADRATIC FORMULA

The method of solution by completing the square is used to solve any form of quadratic equation. We shall now use this method to find the roots of the general quadratic equation, $ax^2 + bc + c = 0$.

THEOREM 11-1:   *The solution set S for equation $ax^2 + bx + c = 0$, with $a \neq 0$, is*

$$S = \left\{ \frac{^-b \pm \sqrt{b^2 - 4ac}}{2a} \right\}$$

Proof: (by completing the square)

$$ax^2 + bx + c = 0$$
$$ax^2 + bx = {}^-c$$

$$x^2 + \frac{b}{a}x = \frac{^-c}{a}, \text{ since } a \neq 0$$

$$x^2 + \frac{b}{a}x + \left(\frac{b}{2a}\right)^2 = \frac{^-c}{a} + \left(\frac{b}{2a}\right)^2$$

$$\left(x + \frac{b}{2a}\right)^2 = \frac{^-c(4a)}{a(4a)} + \frac{b^2}{4a^2} = \frac{^-4ac + b^2}{4a^2}$$

$$\left(x + \frac{b}{2a}\right) = \pm\sqrt{\frac{b^2 - 4ac}{4a^2}} = \frac{\pm\sqrt{b^2 - 4ac}}{2a}$$

$$x = \frac{^-b}{2a} + \frac{\pm\sqrt{b^2 - 4ac}}{2a}$$

$$x = \frac{^-b \pm \sqrt{b^2 - 4ac}}{2a}$$

Therefore, $S = \left\{\dfrac{^-b \pm \sqrt{b^2 - 4ac}}{2a}\right\}$

The quadratic formula gives the two roots in a quadratic equation in terms of the coefficients of x^2, x, and the constant. The a, b, and c in the formula must be taken from a quadratic equation in the general form $ax^2 + bx + c = 0$. An equation like $3x^2 - 2x - 1 = 0$ must be viewed as $3x^2 + {}^-2x + {}^-1 = 0$, where $a = 3$, $b = {}^-2$, and $c = {}^-1$.

examples: (a)   Find S by formula for $x^2 + 3x + 2 = 0$

For this equation, $a = 1$, $b = 3$, and $c = 2$

$$x = \frac{^-b \pm \sqrt{b^2 - 4ac}}{2a} \quad \text{upon substitution becomes}$$

$$x = \frac{^-3 \pm \sqrt{3^2 - 4(1)(2)}}{2(1)} = \frac{^-3 \pm \sqrt{9 - 8}}{2}$$

$$= \frac{^-3 \pm \sqrt{1}}{2} = \frac{^-3 \pm 1}{2}$$

Therefore, $x = \dfrac{^-3 + 1}{2} = \dfrac{^-2}{2} = {}^-1$ or

$$x = \frac{^-3 - 1}{2} = \frac{^-4}{2} = {}^-2$$

$S = \{^-1, {}^-2\}$

(b) Find S if $3x^2 - 5x - 3 = 0$

For this equation, $a = 3$, $b = {}^-5$, and $c = {}^-3$

$$x = \frac{{}^-({}^-5) \pm \sqrt{({}^-5)^2 - 4(3)({}^-3)}}{2(3)}$$

$$= \frac{5 \pm \sqrt{25 + 36}}{6} = \frac{5 \pm \sqrt{61}}{6}$$

$$S = \left\{ \frac{5 \pm \sqrt{61}}{6} \right\}$$

(c) Find S if $2x^2 - 3x = 0$

For this equation, $a = 2$, $b = {}^-3$, and $c = 0$

$$x = \frac{{}^-({}^-3) \pm \sqrt{({}^-3)^2 - 4(2)(0)}}{2(2)}$$

$$x = \frac{3 \pm \sqrt{9 - 0}}{4} = \frac{3 \pm 3}{4}$$

Therefore, $x = \dfrac{3 + 3}{4} = \dfrac{6}{4} = \dfrac{3}{2}$ or

$$x = \frac{3 - 3}{4} = \frac{0}{4} = 0$$

$$S = \left\{ \frac{3}{2}, 0 \right\}$$

Exercise 73

Find S by formula:

1. $y^2 + 3y + 2 = 0$
2. $m^2 + 6m - 7 = 0$
3. $a^2 + 5a - 50 = 0$
4. $x^2 - 12 = 0$ (Hint: $b = 0$)
5. $q^2 - 3q = 0$
6. $2m^2 - m - 1 = 0$
7. $4x^2 + 3x - 1 = 0$
8. $5x^2 - 6x + 1 = 0$
9. $\{x \mid ax^2 + 6x + 4 = 0\}$
10. $\{y \mid py^2 + py + 2 = 0\}$
11. $\{x \mid 3x^2 - 5 = 0\}$
12. $\{m \mid 2m^2 + 6m = 0\}$

13. $\dfrac{1}{2}a^2 - \dfrac{1}{3}a - 1 = 0$ (Clear fractions first.)

14. $\dfrac{2}{3}x^2 - \dfrac{1}{3}x - 3 = 0$

15. Write an equation if $S = \{3, {}^{-}5\}$.

16. *Find S by factoring:* $x^2 + 8x - 33 = 0$

17. *Find S without formula:*
    (a) $2x^2 - 24 = 0$ (b) $2x^2 - 24x = 0$ (c) $x^2 - 1 = 0$

18. If $x = 3 + \sqrt{2}$, evaluate $(x + 3)^2$.

19. *Simplify:*   (a) $2\sqrt{20} - 3\sqrt{45} + 6$
              (b) $3\sqrt{3}\,(\sqrt{3} - 2\sqrt{6})$

Find S by the method which seems best to you:

20. $x^2 - 11x + 10 = 0$ 21. $3x^2 - 9 = 0$

22. $4x^2 + 2x = 0$ 23. $m^3 - 3m^2 = 0$

24. $2x^2 + 7x - 1 = 0$ 25. $4x^2 - 15x - 4 = 0$

26. $3x^2 - 8x + 2 = 0$ 27. $8x^2 - 3x - 5 = 0$

28. Find two numbers whose sum is 120 and whose difference is 44.

29. How many dimes are there in a coin purse that contains 3 times as many dimes as quarters and whose value is $6.05?

30. A statement which we accept as true without proof is called a(n)_____ .

11-7. A BRIEF LOOK AT COMPLEX NUMBERS

Thus far all of the numbers in this book have been real numbers. One property of real numbers is that the square of a real number is always positive or zero.

Suppose the equation $x^2 = {}^{-}1$ is to be solved. No real number can be a root of this equation. To have a solution for $x^2 = {}^{-}1$, let us invent a new number whose square is ${}^{-}1$. This number is to be called "i"; and it will serve to identify a new set of numbers called the *imaginary numbers. Imaginary numbers are numbers whose squares are negative.* For example, $\sqrt{{}^{-}3}$ is imaginary since $(\sqrt{{}^{-}3})^2 = {}^{-}3$.

The following agreements are to be made about all imaginary numbers. First,

$$i = \sqrt{{}^{-}1} \quad \text{and} \quad i^2 = {}^{-}1$$

Second, all imaginary numbers must be written in the "i form" *before any computations are done.* The "i form" is obtained as shown:

    (a) $\sqrt{{}^{-}3} = \sqrt{{}^{-}1(3)} = \sqrt{{}^{-}1} \cdot \sqrt{3} = i\sqrt{3}$

    (b) $\sqrt{{}^{-}4} = \sqrt{{}^{-}1} \cdot \sqrt{4} = i\sqrt{4} = i \cdot 2 = 2i$

The reason for this second agreement is to avoid any ambiguity. Using the "i form," $\sqrt{-3} \cdot \sqrt{-5} = i\sqrt{3} \cdot i\sqrt{5} = i^2\sqrt{15} = (^-1)\sqrt{15}$ $= ^-\sqrt{15}$; but if the "i form" is not used, $\sqrt{-3} \cdot \sqrt{-5} = \sqrt{-3(^-5)} = \sqrt{15}$.

A number which has at least two terms, one of which is real and another is imaginary, is called a complex number.

Some examples of complex numbers are $3 + i\sqrt{2}$, $\dfrac{1+i}{4}$, and $2 + 0i$.

> DEFINITION: *A number of the form $a + bi$, where a and b are real, is a complex number in standard form.*

Complex numbers are added and multiplied according to the previous theorems about real numbers.

examples:

(a) $(2 + 3i) + (5 - 2i) = 2 + 5 + 3i - 2i = 7 + i$

(b) $\sqrt{-12} + \sqrt{-48} = i\sqrt{12} + i\sqrt{48}$
$$= 2i\sqrt{3} + 4i\sqrt{3}$$
$$= 6i\sqrt{3}$$

(c) $2\sqrt{-5} \cdot \sqrt{-3} = 2i\sqrt{5} \cdot i\sqrt{3} = 2i^2\sqrt{15}$
$$= 2(^-1)\sqrt{15} = ^-2\sqrt{15}$$

(d) $(2 + \sqrt{-6})^2 = (2 + i\sqrt{6})^2$
$$= 4 + 4i\sqrt{6} + i^2(\sqrt{6})^2$$
$$= 4 + 4i\sqrt{6} - 6$$
$$= ^-2 + 4i\sqrt{6}$$

(e) Write the number $3x + 2i + 4ix + 6$ in standard complex form:
$$3x + 2i + 4i + 6 = 3x + 6 + 2i + 4ix$$
$$= (3x + 6) + i(2 + 4x)$$

Exercise 74

Perform the indicated operations and write in standard complex form:

1. $3i + 5i$
2. $2i - 3 + 4i$
3. $2i^2 - 3$
4. $(5i)^2$

5. $2(3 + 2i)$ 6. $5i(3 + 3i)$
7. $\sqrt{^-2} - \sqrt{2}$ 8. $2\sqrt{^-2} + i\sqrt{2}$
9. $3 - \sqrt{^-12}$ 10. $1 + \sqrt{^-1}$

11. $5 \pm \sqrt{^-16}$ 12. $\dfrac{3 \pm \sqrt{^-18}}{3}$

13. $\dfrac{25 \pm \sqrt{^-10}}{5}$ 14. $6 \pm 6\sqrt{^-24}$

15. $2 - i\sqrt{18}$ 16. $3i + 2\sqrt{^-36}$
17. $(3 + \sqrt{^-4})^2$ 18. $(\sqrt{3} + 2i)^2$
19. $(\sqrt{5} + 2i\sqrt{5})^2$ 20. $(\sqrt{3} - i)(\sqrt{3} + i)$

Find S:

21. $3x^2 = 9$ 22. $x^2 = 1$
23. $x^2 = 5$ 24. $5x^2 - 16 = 0$
25. $3x^2 + 2x = 0$ 26. $x^2 - 4x + 4 = 0$
27. $9x^2 - 10x - 1 = 0$ 28. $5x^2 + 6x + 1 = 0$

11-8.   QUADRATIC EQUATIONS WITH COMPLEX OR IMAGINARY ROOTS

To solve the equation $x^2 + x + 1 = 0$, we use the quadratic formula, as the polynomial is not factorable. The roots are

$$\frac{^-1 \pm \sqrt{1^2 - 4(1)(1)}}{2} = \frac{^-1 \pm \sqrt{^-3}}{2},$$

which are complex numbers and must be written in standard form,

$$\frac{^-1 \pm i\sqrt{3}}{2} = \frac{^-1}{2} \pm \frac{i\sqrt{3}}{2}.$$

Exercise 75

Find S and simplify roots:

1. $x^2 + 1 = 0$ 2. $x^2 + 4 = 0$
3. $x^2 = {^-16}$ 4. $x^2 + 5 = 0$

5. $a^2 - 3 = 0$ 6. $b^2 - 16 = 0$

7. $(k + 3)^2 = {}^-1$ 8. $(x - 4)^2 = {}^-18$

9. $x^2 + x + 3 = 0$ 10. $2x^2 + x + 5 = 0$

11. $m^2 - 4m = 0$ 12. $b^2 + 4b + 3 = 0$

13. $\{(x,y) \mid (1)\ x + 3y = 8 \quad \text{and} \quad (2)\ 2x - y = {}^-5\}$

14. *Graph* $\{(x,y) \mid 2x + 5y = {}^-10\}$ using the intercept method.

15. $S = \{x \mid 3x - 2(x + 1) + 3 = 4(2x - 1)\}$. Check this root by substitution.

16. *Use the method of substitution:*
     (1) $y = x + 4$
     (2) $x - 3y = 10$

Find S:

17. $3x^2 + 8x + 10 = 0$ 18. $x^2 - 3x = {}^-5$

19. For $px^2 + 2x - k = 0$, are the roots ever imaginary?

20. For $ax^2 + bx + c = 0$, are the roots ever imaginary?

11-9. THE PYTHAGOREAN THEOREM

A classic example of the application of squares, square roots, and quadratic equations is the study of right triangles. A right triangle has one right angle (90 degrees) and two angles less than 90 degrees, called *acute angles.* The side opposite the right angle is called the *hypotenuse.* The other two sides are called the *legs* of the triangle. (See Figure 11-1.)

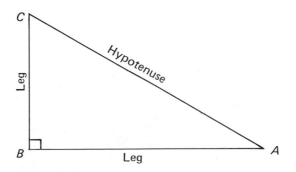

FIGURE 11-1 Triangle *ABC* is a right triangle

THE PYTHAGOREAN THEOREM: *In a right triangle, the sum of the squares of the lengths of the legs is equal to the square of the length of the hypotenuse.*

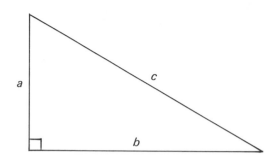

That is, if a and b are the lengths of the legs and c is the length of the hypotenuse,

$$a^2 + b^2 = c^2$$

If the legs of a right triangle are known, the hypotenuse can be found by the above formula.

examples: (a)   If the legs of a right triangle are 6 feet and 8 feet, find the hypotenuse.

Let h = the number of feet in the hypotenuse
$$h^2 = 6^2 + 8^2$$
$$h^2 = 36 + 64 = 100$$
$$h = \sqrt{100} = 10$$

Answer: Hypotenuse of triangle is 10 feet.

(b)   If the hypotenuse of a right triangle is 13 inches and one leg is 5 inches, find the length of the other leg.

Let a = the number of inches in the missing leg
$$13^2 = 5^2 + a^2 \quad \text{or} \quad a^2 = 13^2 - 5^2$$
$$a^2 = 169 - 25$$
$$a^2 = 144$$
$$a = 12$$

Answer: Missing leg is 12 inches long.

(c)   If the point (2,3) is plotted on a rectangular
      coordinate system, how far is it from the origin?

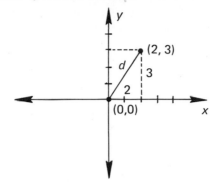

Let d = distance from origin

By the Pythagorean theorem,

$$d^2 = 2^2 + 3^2 = 13$$
$$d = \sqrt{13}$$

Answer: The distance from the origin is $\sqrt{13}$.

DEFINITION: *The diagonal of a rectangle (or a square) is a line segment that is drawn from two opposite vertices. (See Figure 11-2.)*

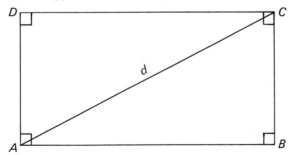

FIGURE 11-2 In the rectangle *ABCD, d* is a diagonal

Exercise 76

If necessary, use the square root table in Appendix B to approximate answers to two decimal places:

1.   Using the Pythagorean theorem $(a^2 + b^2 = c^2)$, find the missing part of each right triangle.

   (a) $a = 1, b = 2, c = ?$

(b) $a = \sqrt{14}$, $b = ?$, $c = 3\sqrt{7}$.

(c) $a = x + 1$, $b = 2x$, $c = ?$

2. A 40-foot guy wire on a telephone pole is anchored 18 feet from the bottom of the pole on level ground. How far up the pole is the wire fastened?

3. A ship sails due east for 16 miles and then due south for 12 miles. At this point how far from its original position has it sailed?

4. If the diagonal of a square is 20 inches, what is the length of each side of the square?

5. A rectangle is two feet longer than it is wide. Its diagonal measures 10 feet. What is the area of the rectangle?

6. A ladder is leaning against a building. The top of the ladder is just touching a window sill 20 feet above the ground, and the foot of the ladder is 10 feet from the building. How long is the ladder?

7. Can a right triangle be made with sides of 7 feet, 24 feet, and 25 feet?

8. A children's slide is advertised as 100 feet long. A rope ladder, 20 feet long, is used to mount the slide. If the slide is 6 feet wide, how much area is there on the ground beneath this toy?

9. Find two numbers whose sum is ⁻80 and whose difference is ⁻16.

10. What is the middle number in a set of three consecutive integers whose sum is 93?

11. Separate $96 into two parts so that one part has $12 more than the other.

12. What is the area of a circle whose circumference is the path of a tetherball at the end of an 8-foot rope? The ball travels around the 10-foot pole at a height 4 feet from the ground.

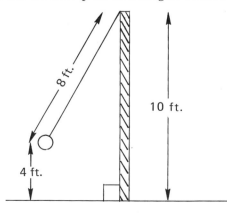

11-10.   STATED PROBLEMS

   Many of the stated problems of this section require a quadratic
equation in their solution. Make certain that the roots you choose
satisfy the conditions of the problem. For example, if x represents
the length of the side of a rectangle and the solution set for the
equation contains 3 and ⁻8, discard the ⁻8, as a rectangle can have
no negative side.

examples: (a)   Find two numbers whose sum is 10 and whose
                      product is 21.

                      Let x = one number
                      $(10 - x)$ = the other number

                      $$x(10 - x) = 21$$
                      $$10x - x^2 = 21$$
                      $$x^2 - 10x + 21 = 0$$
                      $$(x - 7)(x - 3) = 0$$
                      $$x = 7 \quad \text{or} \quad x = 3$$
                      $$S = \{7, 3\}$$

                      Answer: The two numbers are

                      $$x = 7 \qquad x = 3$$
                      $$\text{or}$$
                      $$10 - x = 3 \qquad 10 - x = 7$$

                      Therefore, 7 and 3 is the only pair of numbers.

          (b)   One number is twice another and the square of
                their sum is 324. Find the two numbers.

                      Let x = first number
                           $2x$ = second number

                      $$(x + 2x)^2 = 324$$
                      $$9x^2 = 324$$
                      $$x^2 = 36$$
                      $$x = \pm 6$$
                      $$S = \{\pm 6\}$$

                      Answer: $x = 6$ $x = {}^-6$
                                $2x = 12$ or $2x = {}^-12$

                      For this problem there are two sets of answers.

Exercise 77

For each problem write an equation, solve it, and find the members of S that satisfy the stated problem:

1. Two numbers differ by 5 and their product is 24. Find the numbers.

2. Find two numbers such that the first is 2 more than the second, and the square of the first plus the second (not squared) is 108.

3. If a number is increased by 5, its square is increased by 85. What is the number?

4. Find two numbers that differ by 1 and whose product is 30.

5. If one number is 8 more than twice another and their product is 24, what are the numbers?

6. One leg of a right triangle is 3 feet longer than the other. The hypotenuse is 15 feet. Find the perimeter of the triangle.

7. In a rectangle the area is 40 square inches and the length is 2 less than 3 times the width. What is the length and width?

8. In a square the diagonal is $8\sqrt{2}$ feet. What is the perimeter and area?

9. A square is made into a rectangle by increasing one pair of sides by 4 feet. The area of the rectangle contains 20 square feet more than that of the square. What are the dimensions of the rectangle?

10. If the perimeter of a rectangle is 18 feet and the area is 20 square feet, find the dimensions. (*Hint:* Use two variables and the method of substitution.)

11. One number is 2 more than another number and the sum of their squares is 130. What are the numbers?

12. Paul weighs 160 pounds and his brother weighs 125 pounds. They are sitting on opposite ends of a 16-foot teeter-totter that is balanced. How far from Paul is the fulcrum?

13. Mike has 6 more quarters than dimes, and the rest of his 30 coins are nickels. If he has $4.70, how many nickels does he have?

14. The diagonal of a rectangle is 104 inches and the width is 40 inches. Find the length.

15. What is the area of a piece of land in the shape of a right triangle whose hypotenuse is 52 yards and in which one of the legs is 40 feet?

Appendix A: Five sets of cumulative review problems and sample final examination

Final Cumulative Review I

1. Name five undefined terms used in this course.
2. State the associative axiom for addition and the additive inverse axiom.
3. (a) Give an example of a sum which contains three terms, two of which are similar.

   (b) Give an expression which contains three binomial factors.
4. Subtract $21m^3 - 8m^2 + 3m - 6$ from $50m^3 + 5m^2 - 24$.
5. *Simplify:* $(4x^2 - 3y^2 + 2z^2) - (8x^2 - 6z^2 + 5y^2)$
   $$- (^-3x^2 + 2y^2 - 7y^2)$$
6. (a) Write in set builder notation, "B is the set of all integers between 20 and 26."

   (b) List the members of $D = \{d \mid d$ is a school day $\}$.
7. *Find S:* (a) $\frac{1}{3}(x - 2) + \frac{1}{2} = 6$ (b) $\frac{^-2}{7}x = \frac{3}{8}$
8. *Find S:* $2(x + 3) + 5x + 3(2 - x) = 1 - x$
9. *Find S:* $\{x \mid 3x - 2 \leq 14\}$
10. *Graph:* $\{(x,y) \mid x + y \geq 3\}$
11. *Factor:* $16x^2 y^2 z - 100z^3$
12. *Find S:* $x^2 - 8x + 12 = 0$

13. *Reduce:* (a) $\dfrac{5m^2 - 15m}{m^2 - 9}$ (b) $\dfrac{2m^2 + 5m - 3}{2m^2 - 11m + 5}$

14. What is LCM for $\{21, 24, 30\}$?

15. If $(3,y) \in \{(x,y) \mid 5x - 2y = 6\}$, find y.

16. Find the x- and y-intercepts and graph $3x + 4y = 24$.

17. *Simplify:* (a) $\dfrac{y^{-1}z^2}{y^{-3}z^4}$ (b) $x^{-1} + y^{-1}$

18. *Simplify:* (a) $(3\sqrt{18})(4\sqrt{12})$ (b) $(\tfrac{1}{3}\sqrt{60})(\tfrac{1}{4}\sqrt{27})$

19. Compute $\sqrt{386}$ to three decimal places.

20. *Find S by completing the square:* $x^2 + 8x - 9 = 0$

Final Cumulative Review II

1. (a) Name three irrational numbers other than π.
   (b) Explain how the integers form a subset of the rational numbers.

2. Give the number of terms and factors in each expression.
   (a) $2xy$ (b) $3(x + y)$
   (c) $(3x + 4)(2y + m)$ (d) $2(y - 3) + 4$

3. (a) The word that replaces "addend" is _____ .
   (b) The word for "multiplier" and "multiplicand" is

   _____ .

4. *Subtract* $42x^3m^3 - 8x^2m^2$ *from* $12x^3m^3 + 5x^2m^2$.

5. If $a = 2$ and $b = {}^-1$, evaluate $2a^2b - b^2 + 2b^3 - ab^2$.

6. *Find S:* (a) $\dfrac{3}{4}(x + 3) - \dfrac{1}{2}(2x - 1) = 0$

   (b) $\dfrac{3}{4}(x + 3)(2x - 1) = 0$

7. *Solve for m:*
   (a) $2b = 3m - y$ (b) $am^2 - 2m = 3$ (*Hint:* Use the quadratic formula.)

8. Jeff has \$4.92. He has 4 times as many dimes as pennies. How much are his dimes worth?

9. Write S in set builder notation and graph:
   (a) $3x - 1 \le 14$ (b) $4 - 2x > 21$

10. *Graph:* $\{(x,y) \mid 2x - 5y \ge 10\}$

11. *Factor completely:*
    (a) $12m^3 - 30m^2 - 18m$ (b) $24m^4 + 150m^2$
12. *Multiply:*
    (a) $2x(x + 3)(x - 3)$ (b) $3m(2m + 3)^2$
    (c) $(4m - 1)(6m + 7)$ (d) $3(x + y)(x + y)(x - y)$
13. *Simplify:*

    (a) $\dfrac{3m^2 + 9m}{m^2 - 9} \cdot \dfrac{m^3 - 6m^2 + 9m}{3m^3}$

    (b) $\dfrac{24x^2 y^2}{x^2 - 2xy + y^2} \div \dfrac{8xy^2}{x^2 - xy}$

14. *Find N:*

    (a) $\dfrac{3}{5} = \dfrac{N}{20}$ (b) $\dfrac{^-4x}{3a} = \dfrac{N}{9a^3}$

    (c) $\dfrac{a + 2}{a - 3} = \dfrac{N}{a^3 - 3a^2}$ (d) $\dfrac{1}{xy} = \dfrac{N}{x^2 y^2 - xy}$

15. Given $A = \{(x,y) \mid x - 2y = 6\}$:
    (a) Is $(0,3) \in A$?
    (b) Find y if $(3,y) \in A$.
    (c) What is the x-intercept of the graph of A.
    (d) Is $6 \in A$? Why?
16. Given $\{(x,y) \mid (1)\ x + y = 8$ and $2x - y = 7\}$: Find S by
    graphing carefully and check your solution algebraically.
17. Write in scientific form and simplify if possible:
    (a) 483   (b) .6790
    (c) 0.00361   (d) 0.0184

    (e) $\dfrac{(4800)(90000)(.00012)}{(.024)(180,000)(8)}$ (f) $\dfrac{10^{-3}}{(240)(.0003)}$

18. *Simplify:*
    (a) $\sqrt{600}$ (b) $4\sqrt{250}$ (c) $\sqrt{28x^3}$

    (d) $\sqrt{\dfrac{1}{2}}$ (e) $\sqrt{\dfrac{a}{x}}$ (f) $6x\sqrt{72x^4}$

19. *Find S:*
    (a) $3x^2 = 21$ (b) $\dfrac{1}{2}a^2 = 8$

    (c) $4x^2 - 1 = 0$ (d) $x^2 + 8 = 0$
    (e) $x^2 - 4x = 0$ (f) $9x^2 + 27x = 0$

20. *Solve by quadratic formula:*
(a) $m^2 + 3m - 8 = 0$ (b) $2y^2 - y - 1 = 0$
(c) $5a^2 - 12 = 0$

Final Cumulative Review III

1. Give all the subsets of $A = \{3,4\}$.
2. If $A = \{1,8,a,m\}$ and $B = \{18,p,q\}$, find
(a) $A \cap B$ (b) $B \cup \emptyset$ (c) $A \cup B$ (d) $A \cap \emptyset$
3. *Name the axiom used:*
(a) $ab - 3 = ba - 3$ (b) $(x + 2)(x + 3) = (x + 2)x + (x + 2)3$
(c) $(3a)(bd) = (3ab)d$
4. *Multiply:*
(a) $2x^2 + 3x - 2$ (b) $3x(x + 2)(2x - 5)$
   $4x^2 + 5x - 1$

5. *Find S:*
(a) $9(2 - x) + \dfrac{1}{2}(x - 6) = 1$ (b) $x^2 - 3 = 0$
(c) $2x^2 - 4 = 3$

6. *Solve for m:* $K = 3m - mx$
7. *Find and graph S:* $\left\{ x \mid 3x - \dfrac{1}{2} \leq \dfrac{1}{3} \right\}$
8. *Multiply:*
(a) $(6m + 1)(6m - 1)$ (b) $(6m + 1)^2$
(c) $(16m + 4)(\frac{1}{2}m - 2)$ (d) $(8x^2 - y^2)^2$
9. *Simplify:* $\dfrac{m^2 + 3m}{m^2 - 1} \cdot \dfrac{m^2 + 3m + 2}{m^2 + 2m - 3} \div \dfrac{m}{m^2 - 2m + 1}$
10. *Simplify:*
(a) $\dfrac{1}{m} + \dfrac{2}{m} + \dfrac{3}{m^2}$ (b) $3 - \dfrac{1}{x}$
(c) $a^2 + \dfrac{2}{a^2}$ (d) $\dfrac{y}{z} - \dfrac{3y}{2z} - z$
11. *Add:* $\dfrac{x - 1}{x + 2} + \dfrac{^-3}{x} + \dfrac{x^2}{x^2 + 2x}$
12. *Graph:*
(a) $\{(x,y) \mid 3x = y\}$ (b) $\{(x,y) \mid 2y - x = 4\}$

13. *Find S by addition:*
    (a) (1) $3x + y = 1$
              $2x + y = 0$
    (b) (1) $2x + y = 1 + x$
         (2) $2x - y = 6 + y$

14. *Find S by substitution:*
    (a) $^{-}5x + 2y = 7$
         $y = 4x + 8$
    (b) (1) $x - 3y = 4$
         (2) $2x - y = 0$

15. Green beans sell for 30¢ a can and applesauce for 20¢ a can. If Mrs. Norman spent $3.60 for 14 cans, how many of each did she buy?

16. Paul and Mark sit on opposite ends of a teeter-totter. Paul weighs 110 pounds and is 3 feet closer to the fulcrum than Mark, who weighs 80 pounds. How long is the teeter-totter?

17. *Simplify:*
    (a) $\sqrt{48x^3}$
    (b) $\sqrt{240}$

    (c) $\sqrt{\dfrac{1}{x+1}}$
    (d) $\sqrt{x^2 + y^2}$

18. *Simplify:*
    (a) $3\sqrt{5} - 2\sqrt{5}$
    (b) $2\sqrt{20} - 3\sqrt{75} + 3\sqrt{16}$
    (c) $\sqrt{18x^3} - x\sqrt{50x} - \sqrt{2x^3}$

19. *Find S:*
    (a) $x^2 + 1 = 0$
    (b) $x^2 - 1 = 0$
    (c) $5x^2 = 8$
    (d) $x^2 = 4x$
    (e) $9m^2 = 9m$
    (f) $9m^2 - 9 = 0$

20. *Simplify:*
    (a) $(3 + i)^2$
    (b) $2i(i - 1)$

    (c) $\dfrac{3x^2}{3i}$
    (d) $(5 + 2i)(5 - 2i)$

Final Cumulative Review IV

1. *Simplify:*
    (a) $6 \cdot 3 + 4^2 - 2$
    (b) $6(3 + 4)^2 - 2^2$
    (c) $21 \div 3 \cdot 7$
    (d) $21 \div 7 \cdot 3$

2. *Divide* $8x^3 + 10x^2 - x + 3$ *by* $2x + 3$

3. Find two consecutive even integers such that 5 times the first is 4 times the second.

4. Find four consecutive integers whose sum is 98.

5. *Factor completely:*
   (a) $x^2 + 9x - 10$ (b) $3x^2 + 18x + 27$
   (c) $8b^2 - 59b + 21$ (d) $3x^2 + 75$
6. *Square* $(3x + 2y + 4)$
7. *Simplify:*

   (a) $4 + \dfrac{\dfrac{1}{3}}{2 - \dfrac{1}{2}}$ (b) $1 + \dfrac{\dfrac{1}{x} - \dfrac{1}{x^2}}{2}$

8. *Find S:*

   (a) $\dfrac{3(x-2)}{4} + \dfrac{2(x+1)}{3} - \dfrac{5(x-3)}{6} = 12$

   (b) $3 + \dfrac{x}{4} - \dfrac{x-1}{2} = x - \dfrac{1}{2}$

9. One number is 3 more than twice another number. Their sum is 21. What are the numbers? (Use two variables.)

10. *Find S:*   (1) $\dfrac{y}{2} + \dfrac{x}{3} = {}^{-}\left(\dfrac{1}{3}\right)$

                   (2) $\dfrac{x}{4} - \dfrac{2y}{5} = \dfrac{3}{4}$

11. *Simplify:*

    (a) $\dfrac{3}{\sqrt{2}}$ (b) $\dfrac{12}{\sqrt{6}}$

    (c) $\dfrac{x^2}{\sqrt{x}}$ (d) $\dfrac{xy}{\sqrt{y^3}}$

12. *Simplify:*
    (a) $(\sqrt{3} + 1)(\sqrt{3} - 1)$ (b) $\dfrac{1}{\sqrt{3} + 1}$

    (c) $\dfrac{\sqrt{x}}{1 - \sqrt{x}}$ (d) $\dfrac{6 \pm \sqrt{24}}{6}$

13. *Find S:*
    (a) $8x^2 + 1 = 0$ (b) $x^2 + x + 1 = 0$

14. Evaluate $3x^2 + 4x - 5$ if $x = \dfrac{2 + \sqrt{3}}{3}$

15. A rectangle is twice as long as it is wide. If the length is increased by 11 feet and the width is decreased by 2 feet,

the perimeter of the new rectangle is twice that of the old. Find the dimensions of each rectangle.

16. What is the length of the diagonal of the smaller rectangle in problem 15?

17. *Graph:* $\{(x,y) \mid x \leq y\}$

18. *Simplify:*

(a) $\dfrac{x+1}{x^2-1} + \dfrac{1}{x-1}$　　　　(b) $\dfrac{1+\sqrt{2}}{\sqrt{2}}$

19. *State the axiom used:*

(a) $ax + b = b + ax$　　　　(b) $m(np) = (mn)p$

(c) $ay + 0 = ay$　　　　　　(d) $1 \cdot fg = fg$

20. Name five undefined terms used in this course.

Final Cumulative Review V

1. *Give an example of*
   (a) a difference of two squares
   (b) a quadratic trinomial
   (c) a linear equation in two variables
   (d) a trinomial with a common factor

2. *Give an example of*
   (a) a trinomial that is a square
   (b) a quadratic equation in one variable
   (c) a set whose graph is a line parallel to the x-axis
   (d) a set with no members

3. *Find S (do not use the quadratic formula):*
   (a) $3m^2 - 4m = 0$　　　　(b) $4y^2 - 5y - 6 = 0$

4. *Solve for y:*
   (a) $P = by - y$　　　　(b) $y = P - my + 3y$

5. If $y = \frac{1}{2}$ and $z = \frac{3}{4}$, evaluate

(a) $\dfrac{y+4}{y}$　　　　　　(b) $\dfrac{3yz}{y^2 z^2}$

(c) $\dfrac{3z-y}{z-2y}$　　　　(d) $y(y-z)^2$

6. In a change box there is $4.70 consisting of nickels, dimes, and quarters. There are ⅔ as many dimes as quarters and

1½ times as many nickels as quarters. What is the total number of coins?

7. A rectangle has a length which is 3.6 times the width. If the perimeter is 48 feet, what are the dimensions?

8. *Find:*

   (a) the fourth proportional to 7, 12, and 42

   (b) the mean proportional to 10 and 60

9. Twenty-one rolls of barbed wire are required to fence 4 acres. At the same rate, how many rolls will be needed to fence 30 acres?

10. On a 25-point test, ⅗ of the class received 20 or more points, and ¹⁄₁₀ of the class received from 15 to 19 points. The remaining 9 students received less than 15 points. How many members are in the class?

11. *Solve for c:*

   (a) $a^2 = b^2 + c^2$ (b) $a = c^2 + 2ac^2$

12. A ladder is touching a building at a point 12 feet above the ground. The ladder is 16 feet long. What is the distance from the foot of the ladder to the building?

13. Find two numbers whose sum is 14 and whose product is 40.

14. Find the length and width of a rectangle whose area is 12 square inches and whose perimeter is 16 inches.

15. *Simplify:*

   (a) $\dfrac{21x + 30}{9x + 15}$ (b) $\dfrac{3x^2}{x^3} + 2$

   (c) $(^-3)^{-3}$ (d) $(4a^2 x^3)^3$

16. State three axioms that apply to addition.

17. Solve for x by the quadratic formula: $kx^2 + 2x - p = 0$

18. Four blouses and 5 skirts cost 87 dollars. Three blouses and 2 skirts cost 46 dollars. What would one blouse and one skirt cost?

19. A pair of numbers differ by 4. The sum of their squares is 58. What are the numbers?

20. Find three negative consecutive odd integers such that the sum of their squares is 155.

Sample Final Examination

Indicate true or false:

1. An axiom is never to be proved.
2. $\sqrt{a^2 + b^2} = a + b$
3. If $x = {}^-4$, then ${}^-(-x)^2 = {}^-16$
4. $x^{-1} + y^{-1} = \dfrac{1}{x + y}$
5. $ab = ba$ shows the commutative axiom for multiplication.
6. $a \cdot 1 = a$ is a theorem.
7. ${}^-\sqrt{6}$ is a real number.
8. $({}^-8)^2 \geq 0$
9. $\emptyset$ may be a solution set.
10. $ax^2 + bx + c = 0$ has two roots, a, b, and c, are fixed numbers; and $a \neq 0$
11. $\sqrt{4x^3 y^2} = 2xy\sqrt{xy}$
12. By definition, $a \div b = a\left(\dfrac{1}{b}\right)$, $b \neq 0$
13. $\sqrt{{}^-8}$ is a real number.
14. The graph of a linear equation in two variables is a straight line.
15. $(x - y)(x + y) = x^2 - y^2 = (x - y)^2$
16. $x^{2y + 3} = x^{2y} \cdot x^3$
17. $4y^2 + 16$ is prime.
18. $16x^2 - 24x + 9$ is a perfect square.
19. i^2 is a real number.
20. $\dfrac{x + 1}{x} = 1 + \dfrac{1}{x}$

Factor completely:

21. $a^2 b^2 c - abc^2 + 5ab^2 c$
22. $25r^2 + 30r - 16$
23. $125a^2 - 80$
24. $1 - 16a^4$

Simplify:

25. $\sqrt{\dfrac{1}{x}}$

26. $\dfrac{d^2 - cd}{3d}$

27. $\dfrac{t^2 - 5t}{t^2 - 10t + 25}$

28. $\sqrt{8x^4 y^5}$

29. $\dfrac{1}{x} + \dfrac{1}{x + 1}$

30. $\dfrac{(a - b)^2}{a^2 - b^2}$

31. *Multiply:* $\sqrt{7}\,(4\sqrt{2} - 3\sqrt{7})$

32. Which axiom did you use in problem 31?

33. Solve for K if $M = K - 2(K + 1)$.

34. *Simplify:* $\dfrac{5}{2 - \sqrt{3}}$

35. *Multiply:* $(3x - 7)(4x - 9)$

36. Find the mean proportional to 12 and 16.

37. *Simplify:* $\dfrac{a + \dfrac{1}{a}}{2}$

38. *Find S:* $7x - 3(x - 4) + 2 = 4(x + 6) - x$

39. *Find S:* $3x^2 + 2 = 0$

40. (a) What type of equation is $4x^3 + 2y^3 + 3z^3 = 0$?

    (b) What does the graph of (1) $x + y = 2$ and (2) $x + y = 3$ look like? Do not graph.

41. *Find S by factoring:* $3x^2 + x - 4 = 0$

42. *Find S by substitution:*   (1) $x - 3y = 7$
             (2) $2x + 5y = 1$

43. *Find S by addition:* (1) $2x + 3y = {}^-10$
             (2) $3x + 2y = 0$

44. *Find S:* $\dfrac{x - 1}{2} + 3 = \dfrac{x + 4}{5}$

45. *Graph:* $B = \{(x,y) \mid x \geq 4,\, y \leq 1\}$

46. A 54-foot piece of rope is cut into two pieces so that one piece is 20 feet longer than the other. How long is each piece? (*Hint:* Draw a diagram first. Then write two equations in two unknowns.)

47. A collection of dimes and quarters is worth $3.40. There are 3 times as many quarters as dimes. How many of each are there?

48. How far from F is the 400 pounds? (See diagram.)

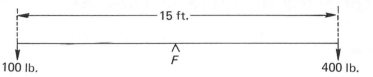

49. Show that there is no *real* number such that the square of the number less 4 times the number is ⁻5. (Assume there is such a real number, etc.)

50. Find x if the shaded area is 438 square feet. (Leave your answer in radical form.)

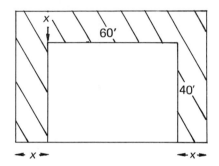

Appendix B: Table of squares, square roots, and prime factors

No.	Sq.	Sq. Rt.	Factors	No.	Sq.	Sq. Rt.	Factors
1	1	1.000		51	2,601	7.141	$3 \cdot 17$
2	4	1.414	2	52	2,704	7.211	$2^2 \cdot 13$
3	9	1.732	3	53	2,809	7.280	53
4	16	2.000	2^2	54	2,916	7.348	$2 \cdot 3^3$
5	25	2.236	5	55	3,025	7.416	$5 \cdot 11$
6	36	2.449	$2 \cdot 3$	56	3,136	7.483	$2^3 \cdot 7$
7	49	2.646	7	57	3,249	7.550	$3 \cdot 19$
8	64	2.828	2^3	58	3,364	7.616	$2 \cdot 29$
9	81	3.000	3^2	59	3,481	7.681	59
10	100	3.162	$2 \cdot 5$	60	3,600	7.746	$2^2 \cdot 3 \cdot 5$
11	121	3.317	11	61	3,721	7.810	61
12	144	3.464	$2^2 \cdot 3$	62	3,844	7.874	$2 \cdot 31$
13	169	3.606	13	63	3,969	7.937	$3^2 \cdot 7$
14	196	3.742	$2 \cdot 7$	64	4,096	8.000	2^6
15	225	3.873	$3 \cdot 5$	65	4,225	8.062	$5 \cdot 13$
16	256	4.000	2^4	66	4,356	8.124	$2 \cdot 3 \cdot 11$
17	289	4.123	17	67	4,489	8.185	67
18	324	4.243	$2 \cdot 3^2$	68	4,624	8.246	$2^2 \cdot 17$
19	361	4.359	19	69	4,761	8.307	$3 \cdot 23$
20	400	4.472	$2^2 \cdot 5$	70	4,900	8.367	$2 \cdot 5 \cdot 7$
21	441	4.583	$3 \cdot 7$	71	5,041	8.426	71
22	484	4.690	$2 \cdot 11$	72	5,184	8.485	$2^3 \cdot 3^2$
23	529	4.796	23	73	5,329	8.544	73
24	576	4.899	$2^3 \cdot 3$	74	5,476	8.602	$2 \cdot 37$
25	625	5.000	5^2	75	5,625	8.660	$3 \cdot 5^2$
26	676	5.099	$2 \cdot 13$	76	5,776	8.718	$2^2 \cdot 19$
27	729	5.196	3^3	77	5,929	8.775	$7 \cdot 11$
28	784	5.292	$2^2 \cdot 7$	78	6,084	8.832	$2 \cdot 3 \cdot 13$
29	841	5.385	29	79	6,241	8.888	79
30	900	5.477	$2 \cdot 3 \cdot 5$	80	6,400	8.944	$2^4 \cdot 5$
31	961	5.568	31	81	6,561	9.000	3^4
32	1,024	5.657	2^5	82	6,724	9.055	$2 \cdot 41$
33	1,089	5.745	$3 \cdot 11$	83	6,889	9.110	83
34	1,156	5.831	$2 \cdot 17$	84	7,056	9.165	$2^2 \cdot 3 \cdot 7$
35	1,225	5.916	$5 \cdot 7$	85	7,225	9.220	$5 \cdot 17$
36	1,296	6.000	$2^2 \cdot 3^2$	86	7,396	9.274	$2 \cdot 43$
37	1,369	6.083	37	87	7,569	9.327	$3 \cdot 29$
38	1,444	6.164	$2 \cdot 19$	88	7,744	9.381	$2^3 \cdot 11$
39	1,521	6.245	$3 \cdot 13$	89	7,921	9.434	89
40	1,600	6.325	$2^3 \cdot 5$	90	8,100	9.487	$2 \cdot 3^2 \cdot 5$
41	1,681	6.403	41	91	8,281	9.539	$7 \cdot 13$
42	1,764	6.481	$2 \cdot 3 \cdot 7$	92	8,464	9.592	$2^2 \cdot 23$
43	1,849	6.557	43	93	8,649	9.644	$3 \cdot 31$
44	1,936	6.633	$2^2 \cdot 11$	94	8,836	9.695	$2 \cdot 47$
45	2,025	6.708	$3^2 \cdot 5$	95	9,025	9.747	$5 \cdot 19$
46	2,116	6.782	$2 \cdot 23$	96	9,216	9.798	$2^5 \cdot 3$
47	2,209	6.856	47	97	9,409	9.849	97
48	2,304	6.928	$2^4 \cdot 3$	98	9,604	9.899	$2 \cdot 7^2$
49	2,401	7.000	7^2	99	9,801	9.950	$3^2 \cdot 11$
50	2,500	7.071	$2 \cdot 5^2$	100	10,000	10.000	$2^2 \cdot 5^2$

Appendix C: Answers to selected problems

Included in this section are the answers to selected odd-numbered problems from the exercises and cumulative reviews presented in the text. Although it seemed useful to provide the answers to most of the odd-numbered problems, certain answers have been omitted based on the following criteria: if the answer is to take the form of a definition or other explanation that could easily be found in the text itself; if more than one answer is acceptable, as in problems that request "an example" of something; if the answer is in the form of a graph, where previous example graphs are sufficiently helpful as a guide to the student. Certain other criteria were used as well, but the three already mentioned were the ones most often employed.

Exercise 2

1. (a) $\{a,b,c,d,1,2,3\}$ (b) $\emptyset$ (c) $\{1,2,3\}$ (d) $\emptyset$
3. $\{1,2,3\}$, $\{1,2\}$, $\{1,3\}$, $\{2,3\}$, $\{1\}$, $\{2\}$, $\{3\}$, $\emptyset$
5. $\dfrac{3}{4}$, $\dfrac{13}{5}$, $\dfrac{1}{6}$, 7
7. A
9. $\{3,7,10\}$
11. $1 \in E$ but $1 \notin F$

Exercise 3

1. Commutative (multiplication)
3. Additive identity
5. Closure (addition)
7. Commutative (multiplication)
9. Associative (addition)
11. Associative (multiplication)
13. Distributive
15. $^-7$
17. (a) 0 (b) 0 (c) 0
19. 3

Exercise 4

1. Term
3. $x - 8$
5. $2x^2$
7. $(x + y)^2$
9. $\dfrac{2x + x}{21}$
11. $\{12,18\}$
13. 2 terms, 1 factor
15. 2 factors, 1 term
17. 3 terms, 1 factor
19. 3 terms, 1 factor
21. #1 is x
    #2 is $2x + 3$
23. $n + 2, n + 4, n + 6$
25. A true statement accepted *without* proof
27. $2n$ has a factor of 2

Exercise 5

1. 5
3. $^-30$
5. 9
7. 10
9. $^-40$
11. $a + b = b + a$
13. $^-9$
15. $^-9$
17. 6
19. 4 terms, 1 factor
21. $^-7$
23. $^-13$
25. 16
27. $^-5$

Exercise 6

1. 2	3. ⁻20
5. 30	7. 70
9. +106	

13. A true statement accepted only with proof

15. 14	17. ⁻39
19. ⁻8	21. ⁻6, 4, $\dfrac{1}{3}$
23. ⁻1	25. 5
27. ⁻15	29. 10
31. 0	

Exercise 7

1. 18	3. ⁻32
5. 6	7. 72
9. 40	11. 500
13. 20	15. 9
17. 64	19. ⁻9
21. 10	23. ⁻30
25. ⁻56	27. $\dfrac{1}{6}$
29. $^{-}\left(\dfrac{1}{24}\right)$	31. Multiplication is commutative
33. ⁻(.006)	35. .0001
37. $\dfrac{49}{4}$	39. .0129

Exercise 8

1. 4	3. ⁻11
5. ⁻6	7. 5
9. 4	11. ⁻73
13. ⁻24	15. 18
17. 9	19. 19

21. To divide a by b means to multiply a by the reciprocal of b.

Exercise 9

1. 30

3. $^-9$

5. $^-\left(\dfrac{1}{2}\right)$

7. 26

9. 21

13. $^-6$

15. $^-4$

17. Integers, rational numbers, irrational numbers

19. 4800

Exercise 10

1. 6

3. $^-7$

5. $^-5$

7. 9

9. $^-2$

11. $^-5$

15. Negative

17. 4

19. $\dfrac{3}{10}$

21. $^-11$

23. Undefined

27. Multiplicative identity

29. First is $(x + y)^2$, second is $x^2 + y^2$

Cumulative Review: Chapters 1 and 2

5. (a) 3   (b) 3   (c) 6

7. $40x^2 y^2$

9. $^-55$

13. (a) $^-63$ (b) 9

Exercise 11

1. First, one, monomial

3. First, two, binomial

5. Third, one, trinomial

7. Second, three, trinomial

9. First, one, binomial

11. Fourth, one, trinomial

13. Similar terms have identical literal coefficients.

15. An axiom is not proven, while a theorem must be proven.

17. $^-51$

19. 42

21. 66

Exercise 12

1. x
5. $9x - 7y$
9. $63x^3y - 3xy$
13. $8a^2 + z^2$
17. 9
21. Zero

3. $^-3k^3z$
7. $12x - 7$
11. $7a - 4b - 3d$
15. $4axy + 1$
19. 15
23. $x + 2k$

Exercise 13

1. $^-3x - 2$
5. $^-3x^2y - 4xy + 3$
9. $^-21m^3 - 6m^2 + 8m + 2$
13. $2c$
17. $^-x + 5$
21. $xy - y - 1$
25. Negative

3. $^-5x + y$
7. ^-xy
11. $^-2x^2 + 6x + 12$
15. $3m^2 + 3m - 2$
19. $^-k^2 - 2k + 6$
23. Integers, rational, irrational

Exercise 14

1. $4x + 4y$
5. $12x^3 - 12x^2 + 12x - 36$
9. $8k^2z + 8kz - 32$
13. $10x - y - 10$
17. 1 term, 2 factors

3. $^-6x + y + 6z$
7. $3a + 2b + 5$
11. $2m^2 + 5m - 38$
15. $8abc + 10ab + 8a$
19. 2 terms, 1 factor

Exercise 15

1. $5x$
5. $^-4a + 20b$
9. $^-x + 18$
13. (a) $^-4$ (b) 24   (c) $^-10$
19. $(x - y - z) - (^-w + u - p)$
23. $m - (3x - 2y)$

3. ^-b
7. $6p - q - 6$
11. Two
17. $(^-xy - 2z + 3) - (7x - 5y + 3)$
21. $6 - (a + b)$
27. { } or $\emptyset$

Exercise 16

1. 9
3. $^-4$
5. $^-22$
7. 72
9. $^-3$
11. $6 + 4 = 4 + 6$
13. Closure for multiplication
15. (a) $5(60 + 5)$ (b) $7(20 + 8)$ (c) $8(300 + 40 + 6)$
        $300 + 25$ $140 + 56$ $2400 + 320 + 48$
        325 196 2768

Exercise 17

1. $6ax^2$
3. $m^2 n$
5. $3a^2 b^2 c$
7. $10k^3$
9. $48x^2 y$
11. $7x^2 + 18x$
13. $8ab^2 - 4a^2 b$
15. $2n^3 p - 5n^2 p + n^2 p^2$
17. $ab = ba$
19. $^-8x^2 y$
21. $6x^2 - 11x - 10$
23. $10m^3 + 2m^2 + 5m + 1$
25. $m^3 + 4m^2 + 4m + 1$
27. $x^3 + 8$
29. $^-3a^4 - 7a^3 - 17a^2 - 3a - 30$
31. $x^2 - 3x - 4$
33. The additive identity
35. 108
37. (a) $^-9$ (b) 9   (c) $^-2$

Exercise 18

1. $36x^3$
3. $^-4$
5. ^-9y
7. $18y^2 - 24y$
9. $7m$
11. $^-60$
13. $\dfrac{^-1}{3}$
15. (a) $20x^2$ (b) $14k$ (c) 19
17. Undefined term
19. $x + 3$
21. $x - 1$
23. $4a - 5$
25. $x - y - \dfrac{6y^2}{x + 2y}$
27. $a^2 + a + 1$

29. $2a^2 + 5a + 5 + \dfrac{^-2a + 3}{a^2 + a - 2}$ 31. $10x^2$

33. $^-8x^2$ 35. $^-2$

Cumulative Review: Chapters 1 through 3

3. $5x + 38$ 5. $12x - 8y$
7. $3x - 9y - 10z$ 9. $4x^2 - 11y^2$
11. (a) $7x^2 - 12xy - 8y^2$ (b) $^-12x^2$
13. (a) $5xy^2 + 5y^2z$ (b) $(5y + 2)\, 3y - (5y + 2)$
15. Linear means first degree; second degree means quadratic.

Exercise 19

1. A statement that two quantities are equal
5. No, $6 + 5 \neq 9$
7. $a(bc) = (ab)c$
9. (a) A = {2,4,6} (b) B = {2,5}   (c) C = $\emptyset$ (d) D = $\emptyset$
    (e) E = {5,7,9,...,17}   (f) F = {Richard Nixon}
    (g) G = {0}   (h) H = $\emptyset$
11. $^-30$
13. Integers, rational, irrational
15. A set which contains the root(s) of an equation
17. $12x^4 + 6x^3$
19. {4}

Exercise 20

1. {5}   3. {0}
5. {2}   7. {2}
9. {$^-12$}   13. $^-12$
15. Third degree, two variables   17. {$^-15$}
19. {$^-2$}   21. {$^-7$}
23. {20}   25. {45}

Exercise 21

1. $\{3\}$

5. $\{4\}$

3. $\{^-19\}$

7. $\{18\}$

9. $\left\{\dfrac{1}{30}\right\}$

11. $^-\left(\dfrac{4}{3}\right)$

13. Additive identity

17. $\{24\}$

15. $\{0\}$

19. $\{^-3\}$

23. $x^2 + y^2$

25. $\left\{\dfrac{7}{54}\right\}$

Exercise 22

1. $\{15\}$

5. $\{^-3\}$

3. $\{^-2\}$

7. $\{1\}$

9. $\left\{\dfrac{5}{2}\right\}$

11. $3x^3 - x^2 - 3x - 8$

13. ^-2a

17. $\{^-43\}$

23. $\{^-2\}$

27. One

15. $a(b + c) = ab + ac$

19. $\{3\}$

25. $\{4\}$

29. $\{8\}$

31. $\left\{\dfrac{10}{9}\right\}$

33. $\left\{\dfrac{1}{8}\right\}$

35. $\{^-2\}$

37. $\left\{^-\left(\dfrac{1}{2}\right)\right\}$

39. $\left\{^-\left(\dfrac{1}{25}\right)\right\}$

41. $\{0\}$

43. $\left\{\dfrac{1}{200}\right\}$ or $\{.005\}$

Exercise 23

1. $s = \dfrac{P}{4}$

3. $r = \dfrac{C}{2\pi}$

5. $v_1 = V - v_2$

7. $h = \dfrac{V}{lw}$

9. $r^2 = \dfrac{S}{4\pi}$

11. $F = \dfrac{9c}{5} + 32$

13. $a = P - b - c$

15. $v = \dfrac{Q}{m}$

17. \$300

19. $\left\{\dfrac{13}{2}\right\}$

21. $\left\{\dfrac{1}{3}\right\}$

23. $\{^-5\}$

25. $\left\{\dfrac{7}{20}\right\}$

27. $\emptyset$

Exercise 24

1. (a) $x, 3x$ (b) $3x + 4$ (c) $x + 3x$ (d) $2x + 3x$
3. (a) $n + d + p$ (b) $5n$ (c) $5n + 10d$ (d) $5n + 10d + p$
5. $25k$
7. $x - 7$
9. $7 - x$
11. $.75x$
13. (a) x dimes, $2x$ nickels (b) $5(2x)$ (c) $10x$ (d) $5(2x) + 10x$
15. (c) $x + (x + 1) = 53$ (e) 26, 27
17. (c) $x + (x + 2) + (x + 4) = 93$ (e) 29, 31, 33
19. (c) $x + (x + 8) = 24$ (e) 8 ft., 16 ft.
21. (c) $2(w + 80) + 2w = 500$ (e) $l = 165$ ft., $w = 85$ ft.
23. (c) $x + (x + 2124) = 7486$ (e) Camper \$2681, truck \$4805
25. (c) $10x + 5(x + 8) = 115$ (e) 13 nickels
27. (c) $x + (5x + 3) = {}^-21$ (e) $^-17$
29. (c) $250x + 150(600 - x) = 112,000$
    (e) 380 children, 110 adult couples
31. (c) $.055x + .04(1200 - x) = 60$ (e) \$800
33. (c) $2x + (2x + 4) + x = 39$ (e) 14, 18, 7
35. (c) $x + 8 = 2(x + 2)$ (e) Anne is 4 years, John is 10 years
37. (c) $300 = .28x$ (e) $1071^3/_7$ pounds
39. (c) $.08x = .06(700 - x) + 3$ (e) At 8% \$321.43, at 6% \$378.57

Cumulative Review: Chapters 1 through 4

3. Negative
5. A set which contains the root or roots of an equation
7. $6a^2 - 5a + 5$ 9. 38
11. $^-(^-3x^2 + 4x + 5)$ 13. (a) 32   (b) 2   (c) $^-4$
15. Real numbers

Exercise 25

1. $3 < 8$
3. (a) {1,2,3,4}   (b) Not possible   (c) {7,8,9,...}
5. (a) No   (b) Yes

Exercise 26

1. $\{x \mid x < ^-3\}$ 3. $\left\{x \mid x > \dfrac{7}{2}\right\}$

5. $\{x \mid x \leq ^-4\}$ 7. $\{x \mid x \geq 5\}$

9. $\{x \mid x < ^-1\}$ 11. $\left\{x \mid x > \dfrac{^-5}{2}\right\}$

13. $\{x \mid x < 0\}$ 15. $\left\{x \mid x < \dfrac{^-1}{4}\right\}$

17.
       $^-3$

19.
       $\dfrac{1}{4}$ 7

Exercise 27

1. $2 \cdot 5$ 3. $3 \cdot 3 \cdot 5$
5. $2 \cdot 2 \cdot 2 \cdot 2 \cdot 3$ 7. $^-2 \cdot 2 \cdot 2$
9. $5 \cdot 5 \cdot 5$ 11. $2 \cdot 2 \cdot 2 \cdot 11$
13. $2 \cdot 3 \cdot 13$ 15. $2 \cdot 2 \cdot 2 \cdot 5 \cdot 5 \cdot 5$
17. $2 \cdot 7^2$ 19. $2 \cdot 3 \cdot 5^3$

21. $\left\{^-\left(\dfrac{1}{2}\right)\right\}$ 23. $\left\{^-\left(\dfrac{11}{2}\right)\right\}$

Exercise 28

1. $3a + 3b$ 3. $2ax + bx$
5. $2p^2 + 8$ 7. $^-3a - 15$
9. $3a^2 + 15a$ 11. $^-k^2 + 2k$
13. $4m^2 + 12m$ 15. $2m^2 + 2mn - 3m$
17. $\{^-5\}$ 19. One
21. Add equal quantities to each side of an equation.
23. $axy + axz + 2ax$ 25. $12k^3 + 18k^2 + 6k$
29. $a^3bx + a^2b^2x + 3abx$ 31. $p^4q^2 + 3p^3q^3 - 4p^2q^4 - 6pq^5$

Exercise 29

1. $2(a + b)$ 3. $5(a - 2b)$
5. $11(m - 3)$ 7. $5x(1 + 2y)$
9. Prime   11. $3m(3m - 4)$
13. $25x^3 + 10x^2$ 15. $3k(5k - 1)$
17. A process of finding prime factors
19. $a(x^2 - y^2)$ 21. $2(4t^2 - 1)$
23. Prime   25. $2(8x^3 + 10x^2 - 7x - 6)$
27. Prime   29. $P(1 + rt)$

31. $P = \dfrac{A}{1 + rt}$ 33. $h = \dfrac{V}{2g}$

35. $m = \dfrac{S}{2v_1 + 3v_2 + 2v_3}$ 37. $g = \dfrac{T^2}{2h_1 + 2h_2}$

39. $x_2 = \dfrac{D - Kx_1}{^-K}$ or $\dfrac{Kx_1 - D}{K}$

Exercise 30

1. $x^2 - 25$ 3. $a^2b^2 - 1$
5. 399 7. $36a^2 - 25$
9. $p^2q^2r^2 - 9$
11. No. Its factors are not identical.
13. $6a^2 + 8ab$ 15. $3x + 1$
17. $pq(q - 1)$ 19. $33, 35$

21. $9x^2 - 1$

23. $4m^2n^2 - 121$

25. $P = \dfrac{A}{1 + rt}$

27. $y = \dfrac{2x}{1 - k}$

29. 1599

31. 2491

33. 39,936

35. 9999

Exercise 31

1. $(x + 6)(x - 6)$
5. $(a + 1)(a - 1)$
9. $(2x + 5)(2x - 5)$
13. $(2x + 7)(2x - 7)$
19. 18

3. $(3x + 1)(3x - 1)$
7. $2(2x + 1)(2x - 1)$
11. $3(x^2 - 3)$
15. $ax(x + 1)(x - 1)$

21. (a) $\left\{\dfrac{35}{3}\right\}$ (b) $\{^-8\}$ (c) $\{0\}$ (d) $\{0\}$ (e) $\left\{\dfrac{4}{3}\right\}$

23. $M = \dfrac{w - 1}{^-z - y}$ or $\dfrac{1 - w}{z + y}$

29. $3(x + 1)(x - 1)$

31. $8(x + 1)(x - 1)$
35. $6t(t + 1)(t - 1)$
39. Prime

33. $4x(y^2 - 3)$
37. $x(a + p + q + b)$
41. $(6x + 1)(6x - 1)$

Exercise 32

1. $x^2 + 3x + 2$
5. $x^2 - 6x - 16$
9. $2x^2 - 3x - 35$
13. $x^2 - 10x$
17. $2(x^2 - 2)$
21. Prime
25. $9x^2 - 6x + 1$
29. $x^3 + 12x^2 + 47x + 60$
33. $ab = ba$
37. $6a^2 - 23a - 4$
41. $x^2 + 6x + 9$
45. $9x^2 - 6x + 1$
49. $x^2 - 14x + 49$

3. $x^2 + 7x + 12$
7. $9x^2 - 9x + 2$
11. $3x^2 + 12x$
15. $18 - 39x + 20x^2$
19. $7x(3x^2 - x - 1)$
23. $ax(x - 1)$
27. $9x^2 - 15x - 6$
31. $\{3\}$
35. For example, $^-1, ^-2, ^-3$
39. $x(a + x)(a - x)$
43. $x^2 - 16x + 64$
47. $25x^2 + 20x + 4$

Exercise 33

1. $x^2 + 2x + 1$

3. $x^2 + 14x + 49$

5. $4x^2 + 4x + 1$

7. $2500 + 100 + 1 = 2601$

9. $x^2 - 9$

11. $t^2 + 10t + 21$

13. $a^2 + 2ab + b^2$

15. $9x^2 - 30x + 25$

17. $x^4 - 72x^2 + 1296$

19. $m(a + b + c)$

21. $3(x + 3)(x - 3)$

23. $abx(x - 1)$

25. A set which contains the root(s) to an equation

27. 8 dimes

29. $4x^2 + 28x + 49$

31. $n = 1$

33. $n = 4$

35. $n = \dfrac{7}{2}$

37. $n = 10$

Exercise 34

1. $(x + 3)(x + 1)$

3. $(x - 3)(x - 1)$

5. $(x - 5)(x + 2)$

7. $2a(x + 2)(x - 2)$

9. $(5 + 4k)(5 - 4k)$

11. $p(qr + rx + qx)$

13. $(k - 7)(k - 4)$

15. $(6x - 1)(x - 4)$

21. $2(4x + 3)(x - 3)$

23. $(a - 5)(a + 1)$

25. $(x - 7)^2$

27. $(8 + a)(1 - a)$

29. $(b + 8)(b - 3)$

33. $(2x + 3)(x - 4)$

35. Prime

37. $\{3\}$

39. $(4a^2 + 25)(2a + 5)(2a - 5)$

41. $9(4y^2 + z)(4y^2 - z)$

43. $3(5a + 3b)(2a + b)$

45. $z^2(4xy + 1)(4xy - 1)$

47. $(t + 7)(t - 4)$

49. $(9m - 1)(m - 18)$

Exercise 36

1. $a^2 + b^2 + 1 + 2ab + 2a + 2b$

3. $a^2 + b^2 + 9 - 2ab - 6a + 6b$

5. $25m^2 + p^2 + 16q^2 + 10mp + 40mq + 8pq$

7. $25p^4 + 9 - 20p^3 - 26p^2 + 12p + 4p^2$

Exercise 37

1. Polynomial with a common factor
3. Polynomial with a common factor
5. A trinomial square
7. A quadratic trinomial with a common factor
9. None of the above
11. A trinomial square with a common factor
13. The product of two binomials that contain similar terms
15. The square of a trinomial   17. $3(2x + 5)(2x - 5)$

19. $9(y + 2)^2$ 21. $\left\{\dfrac{2}{9}\right\}$

23. $\{40\}$ 25. Prime

Exercise 38

1. $\{^-2, 2\}$ 3. $\{0, 4\}$

5. $\{^-9, 1\}$ 7. $\left\{^-\left(\dfrac{2}{5}\right), \dfrac{1}{2}\right\}$

9. $\{0, 4, 3\}$ 11. $\left\{\dfrac{4}{3}\right\}$

13. $\{0, 2\}$ 15. $\left\{\dfrac{15}{2}\right\}$

17. $\{^-5, 5\}$ 19. $\left\{\dfrac{5}{2}, 3\right\}$

23. The factors are not the same.   27. $9x^2 - 25$
29. $6x^2 - 5x - 25$ 31. $6x^3y + 4x^2y^2 + 2xy^3$

33. 28 and 32   35. $\left\{\dfrac{2}{3}\right\}$

Cumulative Review: Chapters 1 through 6

1. 10
3. $^-5$

5. (a) $(x + 2y)(x - 2y)$ (b) $(x - 4)(x + 1)$
    (c) $5(x^2 - 10)$ (d) $2(x + 1)^2$

7. $(2x - 3)(x + 2)$

9. $41, 42, 43$

13. (a) $x^2 - y^2$ (b) $6x^2 - 7xy - 3y^2$
    (c) $4x^2 + 12x + 9$ (d) $3x^2 - 6x + 3$

15. 33 nickels

17. Middle term should be negative

21. $\{0, 3\}$ 23. $\{^-10, ^-1\}$

25. $\{^-5, 5\}$ 27. $x = \dfrac{A}{p + q}$

Exercise 39

1. $\dfrac{^-2}{3}$ 3. $\dfrac{^-2}{x}$

5. $\dfrac{3}{x + 2}$ 7. $5\left(\dfrac{1}{6}\right)$

9. $x\left(\dfrac{1}{x + 2}\right)$ 11. $^-x\left(\dfrac{1}{x - 1}\right)$

13. $\dfrac{3}{7}$ 15. $\dfrac{x + 2}{x}$

17. $\dfrac{axy}{3m}$

Exercise 40

1. $\dfrac{4}{5}$ 3. $\dfrac{8}{9}$

5. 3 7. $4ky$

9. $\dfrac{7p^2 z^3}{4}$ 11. $\dfrac{1}{2}$

13. $x + 4$ 15. $x + 1$

17. $\dfrac{y}{x + 1}$

19. $\dfrac{3}{4t}$

21. $\dfrac{x^2 + 4}{x^2 - 4}$

23. $x^2 + 3$

25. The multiplicative identity

31. Prime

33. (a) $25y^2 - 40y + 16$ (b) $25y^2 - 16$

35. (a) y (b) $\dfrac{1}{a + 1}$

37. $2m^2 + 3m - 1$

39. $\dfrac{m}{k + y}$

41. $\dfrac{x - 1}{4x}$

43. $(a^2 + 9)(a + 3)$

47. $\dfrac{a + 5}{a - 3}$

49. $\dfrac{a + 3}{3(a - 3)}$

Exercise 41

1. $\dfrac{15ax}{14}$

3. $\dfrac{5m}{7n}$

5. $2a$

7. $\dfrac{a}{a + 1}$

9. $\dfrac{y - 2}{y}$

11. $\dfrac{4}{p - 4}$

13. $3m^2$

15. $\dfrac{(z - 3)^2}{(z + 4)^2}$ or $\dfrac{z^2 - 6z + 9}{z^2 + 8z + 16}$

17. $\dfrac{y + 3}{y + 4}$

19. $\left\{\dfrac{-3}{2}\right\}$

21. Distributive

23. (a) 1 term, 3 factors   (b) 2 terms, 1 factor   (c) 2 terms, 1 factor

25. $\dfrac{2r^2 + rs}{r + s}$

27. $\dfrac{3p}{2}$

29. $\dfrac{(2p+1)(p-3)(p+4)}{(3p+2)(p-4)(p+3)}$

31. (a) $g = \dfrac{2k}{t^2}$ (b) $g = \dfrac{^{-}t}{m-1}$ or $\dfrac{t}{1-m}$

33. 13, 14, 15

Exercise 42

1. $\dfrac{2}{3}$ 3. $\dfrac{pq^2}{a^2}$

5. $\dfrac{x^2 + 2x + 1}{x}$ 7. $\dfrac{c^4}{a^4 b}$

9. $\dfrac{10y^2(2y+1)}{(2y-1)(y+2)}$ 11. $\{3\}$

13. $\dfrac{x^2 - 2x - 3}{xy + y^2}$ 15. Zero

Exercise 43

1. 15   3. 450
5. $8x$ 7. $2ab$
9. $(a-1)(a-2)$ 11. $12a(a^2 - 1)$

15. $\dfrac{x-1}{x+1}$ 17. $abcd^2$

19. $y^2(y+3)$ 21. $(m+2)(m+3)(m-3)(m+1)$
23. $x^2 - 8x$ 25. $(p-3)(p+2)(p+5)(p-3)$

Exercise 44

1. $\dfrac{20}{24}$ 3. $\dfrac{18xy}{27x^2}$

5. $\dfrac{x^2 + 3x + 2}{x^2 - 4}$ 7. $\dfrac{15px^2y}{25p^2q}$

9. $\dfrac{18}{6}$ 11. (a) $\dfrac{a}{b}$ (b) $\dfrac{3x + 12}{x + 1}$

13. (a) $9x^2 - 6x + 1$ (b) $25x^2 - 4$
    (c) $6ax - 3x + 14a - 7$ (d) $12a^3 - 60a^2 + 75a$

15. (a) $\dfrac{x^2 + 2x}{x^3 - 4x}$ (b) $\dfrac{2a^2 - 2a}{a^3 - a}$

17. $a(b + c) = ab + ac$
19. (a) $^-1$ (b) 81
21. (a) $am(m + 1)(m - 1)$ (b) $p(q - 2)^2$

Exercise 45

1. $3x^2(9x^2 - 1)$ 3. $\dfrac{7}{a}$

5. $\dfrac{5a + 1}{a + 1}$ 7. $x - 3$

9. $\dfrac{2y - 1}{y}$ 11. $\dfrac{^-5x - 26}{(x - 3)(x - 2)}$

13. (a) $6p^2 + 19p - 7$ (b) $4p^2 + 28p + 49$
    (c) $4x^3 - 36x$ (d) $5m^3 + 35m^2 + 60m$

15. $\dfrac{a^2 + 2a + 3}{3}$

17. (a) $N = 2xy$ (b) $N = (x + 5)(x + 2)$

19. $\dfrac{^-2x}{15}$ 21. $\dfrac{3y + 2}{3y}$

23. $\dfrac{ad + bc}{bd}$ 25. $\dfrac{pq + p}{q}$

27. $\dfrac{ab^2 + a^2}{b^2}$ 29. $\dfrac{y^2 - 4y - 3}{3y}$

31. $\dfrac{x^2 + 2x + 2}{x + 1}$ 33. $\dfrac{x^2 + x + 1}{x^3}$

35. $\dfrac{p - q + 1}{p^2 - pq}$ 37. $\dfrac{5xy - y^2}{x^2 - y^2}$

39. $x(x + y) = (x + y)x$

41. $\dfrac{^-1}{2},\ 2$

43. $\dfrac{^-12a - 8}{3a^2 + 16a + 16}$

45. $\dfrac{p^2 + 3p + 6}{2p^2 + 10p}$

47. $\dfrac{3x^2 - 8x - 10}{x^2 - 3x - 4}$

49. $\dfrac{11x + 13}{(2x - 3)(3x + 4)(x + 1)}$

Exercise 46

1. $\dfrac{6}{7}$

3. $\dfrac{6}{5}$

5. $\dfrac{a^3}{b^3}$

7. $\dfrac{b + 1}{2b + 1}$

9. $\dfrac{a^2 + b}{2a^2 + b}$

11. $\dfrac{1}{2}$

13. $\dfrac{11}{2x}$

15. (a) $\dfrac{2x + 1}{3}$ (b) $\dfrac{a}{b + 1}$

17. $\left\{\dfrac{2}{3},\ \dfrac{^-5}{2}\right\}$

19. $\dfrac{6a + a^2}{8 + 6a^2}$

21. $\dfrac{1}{x + 1}$

23. $2\dfrac{1}{2}$ or $\dfrac{5}{2}$

Exercise 47

1. $\dfrac{^-9}{4}$

3. $\dfrac{^-1}{6}$

5. $\dfrac{^-19}{8}$

7. $\dfrac{2}{3}$

9. Undefined

11. (a) $\dfrac{5}{6a}$ (b) $\dfrac{3x^2y - 2x^2 - 6}{6x}$

13. $\dfrac{81}{2}$

15. $\dfrac{68}{7}$

17. 5

19. $\dfrac{65}{16}$

Exercise 48

1. $\{15\}$

3. $\left\{\dfrac{5}{2}\right\}$

5. $\left\{\dfrac{1}{6}\right\}$

7. $\{^-6\}$

9. $\{^-5\}$

11. $\{4\}$

13. $\{^-2, 1\}$

15. (a) $\dfrac{8pq + 1}{8p - 64}$ (b) $\dfrac{x^3 - x}{x^3 - 1}$

17. (a) $\dfrac{1}{r + 3}$ (b) $\dfrac{3}{4}$

19. $\{4\}$

21. $\{^-19\}$

23. $\left\{\dfrac{^-62}{11}\right\}$

25. $\emptyset$

29. $\{1\}$

31. $\left\{\dfrac{5}{11}\right\}$

33. $\{27\}$

35. $\{0\}$

Exercise 49

1. 28

3. 45 dimes

5. 16, 24

7. 36,000 5¢ bars
   27,000 8¢ bars

9. (a) x (b) $\dfrac{b}{a + 1}$ (c) $\dfrac{a + 1}{a + 3}$ (d) $\dfrac{mx - 2x}{a}$

11. 60 pounds

13. One foot from the 160-pound man

15. (a) $ab^2(a^2 - a - b)$ (b) $(4x + 3)(4x - 3)$
    (c) $(3m + 1)(m - 1)$ (d) $a(x + y)(x - y)$

17. $^-6$

19. 2200 pounds

Exercise 50

1. (a) $\dfrac{3}{2}$ (b) $\dfrac{2}{1}$

   (c) $\dfrac{3}{4}$ (d) $\dfrac{77}{3}$

   (e) $\dfrac{2}{3}$ (f) $\dfrac{2}{3}$

   (g) $\dfrac{5}{1}$ (h) $\dfrac{10}{1}$

   (i) $\dfrac{1}{9}$ (j) $\dfrac{10}{3}$

3. (a) {7}   (b) {6}

   (c) $\left\{1\dfrac{1}{7}\right\}$ (d) $\left\{\dfrac{5}{6}\right\}$

   (e) {3}   (f) {6 or ⁻6}

   (g) {8}   (h) {1.6}

   (i) $\left\{\dfrac{^-1}{3}\right\}$ (j) {9}

   (k) $\left\{\dfrac{7}{2}\right\}$ (l) {3}

5. 63

Exercise 51

1. 128 cups of pineapple juice, 96 cups of orange juice, 16 cups of lemon juice
3. $3000, $2400, $1800, $1200, $600
5. 35 days
7. 16 inches
9. 10 miles
11. (a) 45 lemons   (b) $7.65

Cumulative Review: Chapters 1 through 7

3. (a) $\dfrac{3m + 2}{5m - 25}$ (b) $\dfrac{1}{3}$

7. (a) $\dfrac{9a - 2}{12a}$ (b) $\dfrac{7x^2 + 11x + 2}{6x}$

9. (a) {⁻4, ⁻1}   (b) {0, 3}

15. $2a - 35$

17. The roots of an equation

19. 96 bon-bons

Exercise 52

1. (b) No   (c) Yes
    (d) $y = 0$ (e) $y = 7$
    (f) $x = 5$ (g) $x = {}^-3$

5. (a) $\left\{\dfrac{9}{4}\right\}$ (b) $\{0, 2\}$

    (c) $\{3\}$ (d) $\left\{\dfrac{7}{3}\right\}$

7. (a) $\dfrac{1 + 2x}{x}$ (b) $\dfrac{3a + 4}{4a}$

    (c) $\dfrac{a^2 x^4 + a^2 y^3}{xy}$

9. (a) No   (b) Yes
11. (a) $(a + 7)(a - 1)$ (b) $2a(x - 1)^2$

Exercise 53

15. Yes, since $2(3) - 4 = 2$
19. (a), (b), (c), (d): for each answer you will get one point which is the point of intersection of the two lines.
21. No, 3 is not an ordered pair
23. Yes

Exercise 54

17. $y = 3$
19. x-intercept $= {}^-6$; y-intercept $= 3$
23. $\{4\}$

25. (a) $\dfrac{x}{x + 3}$ (b) $\dfrac{1}{a - b}$ (c) $\dfrac{4x - 3}{20x - 5}$

27. $\dfrac{2}{x}$

Exercise 55

11. Inconsistent
13. 3
15. x-intercept = $\dfrac{9}{4}$; y-intercept = 6

21. (a) $\left\{2, \dfrac{^-3}{2}\right\}$ (b) $\{0, 1\}$

29. (a) $y = \dfrac{k}{5}$ (b) $y = 2a^2 + 2a + 8$

Exercise 56

1. $\{(3, ^-1)\}$
5. $\{(6, ^-14)\}$
9. $\{(^-10, ^-1)\}$
13. $\{(2, 4)\}$

3. $\{(4, 3)\}$
7. $\{(^-7, 4)\}$
11. $\{(1, 1)\}$
15. (a) $9x^2$ (b) $9x^2 - 12x + 4$

17. $x = ^-2, y = \dfrac{2}{3}$

19. $\{(^-5, 3)\}$

21. $\left\{(1, \dfrac{1}{3})\right\}$

23. $\{(3, 0)\}$

25. Middle term should be ^-42x
31. $\{(3, 1)\}$

27. $k = ^-17$

Exercise 57

1. $\{(4, 2)\}$

3. $\{(8, 13)\}$

5. $\{(2, 1)\}$

7. $S = \left\{\left(\dfrac{^-32}{3}, \dfrac{^-25}{3}\right)\right\}$

9. $\{(5, 11)\}$

11. $\{(13, 4)\}$

13. $\{4, ^-2\}$

15. (a) $\left\{(2, \dfrac{5}{2})\right\}$ (b) $\{(0, 6)\}$

17. $(30, {}^-6)$

19. $\left\{\left(\dfrac{{}^-7a - 3}{2}, \dfrac{{}^-3a - 1}{2}\right)\right\}$

21. $S = \left\{\left(\dfrac{{}^-1}{14}, \dfrac{{}^-3}{14}\right)\right\}$

25. (a) $\dfrac{1 + 4x}{x}$ (b) $\dfrac{{}^-a - 5b}{a^2 - b^2}$

   (c) $\dfrac{x^2 + x + 1}{x^2}$ (d) $\dfrac{10ax^2 + 3}{x}$

29. $\left\{\left(\dfrac{{}^-6}{5}, \dfrac{38}{5}\right)\right\}$

Exercise 58

1. $\{({}^-3, {}^-1)\}$

3. $\left\{\left(\dfrac{{}^-5}{8}, \dfrac{{}^-15}{8}\right)\right\}$

5. $\left\{\left(\dfrac{45}{32}, \dfrac{5}{16}\right)\right\}$

7. $\{(0, {}^-1)\}$

9. $\left\{\left(\dfrac{c_1 b_2 - c_2 b_1}{a_1 b_2 - a_2 b_1}, \dfrac{a_1 c_2 - a_2 c_1}{a_1 b_2 - a_2 b_1}\right)\right\}$

11. $\left\{\dfrac{1}{9}\right\}$

13. $\{0, 3\}$

15. $\left\{\left(\dfrac{{}^-20}{3}, \dfrac{38}{3}\right)\right\}$

Exercise 59

1. 25, 65
3. 4, 7
5. Car costs $4300; boat costs $1300
7. Length = 95 feet; width = 55 feet
9. $8\,{}^4/_7$ feet
11. 1971: 7.25 inches; 1972: 7.55 inches
13. 200 dimes
15. $5\,{}^1/_3$ feet
17. $3600 at $7\,{}^1/_2$%; $2400 at 6%
19. $1920

Cumulative Review: Chapters 1 through 8

3. $3x^2 - 7x + 18 - \dfrac{38}{x + 3}$

5. (a) $\{5\}$ (b) $\left\{\dfrac{20}{3}\right\}$

    (c) $\left\{0, \dfrac{4}{3}\right\}$ (d) $\{^-1, 5\}$

7. $\{(5, 6)\}$

9. (a) $\dfrac{2 + x}{2x}$ (b) $\dfrac{4 + x}{6x}$

    (c) $\dfrac{3x + 1}{x}$ (d) $\dfrac{1}{a + b}$

11. Distributive
13. Yes, one
15. 4 cards of pins, 6 spools of thread

17. (a) $\dfrac{3}{2}$ (b) $\dfrac{4 + 4k}{k^2 x}$

19. $\{24\}$

Exercise 61

1. x^5

5. y^6

9. $3a^6$

13. $p^8 q^4$

17. $\dfrac{2}{25}$

21. $a + b$

25. $\dfrac{9}{2}$

3. $6a^3 b^3$

7. x^{k+3}

11. $\dfrac{1}{p^5}$

15. $\dfrac{1}{q^4}$

19. 3

23. $125x^3 y^6$

27. (a) $\{^-3, 2\}$ (b) $\{1, 2\}$

29. $(x + y)^6$

31. $\dfrac{a^6 b^6}{c^3}$

33. y^6

35. $64x^6 y^6 z^4$

37. $\{(7, 2)\}$

39. $\{^-1, 1\}$

41. $50a^6$

43. x^{k+6}

45. $32a^5$

47. $\dfrac{27}{125}$

49. $\dfrac{1}{x^2 + 6x + 9}$

51. $\dfrac{1}{x^2 + 8x + 16}$

53. $\dfrac{64x^3}{y^3}$

55. $9, 11$

Exercise 62

1. $\dfrac{1}{5x}$

3. $\dfrac{a}{b^3}$

5. $\dfrac{yz^3}{6}$

7. $\dfrac{3}{a^2 b^2}$

9. $\dfrac{1}{p^3 q^4}$

11. $\dfrac{1}{10^6}$

13. $\dfrac{1}{2y}$

15. $\dfrac{x + 1}{x}$

17. $\dfrac{1}{p^9 q^9}$

19. $\dfrac{1}{x}$

21. $6x^2$

23. 10^8

25. 1

27. 10^{11}

29. $\dfrac{1}{525}$

31. $\dfrac{a + b}{ab}$

33. $\dfrac{11}{18}$

Exercise 63

1. 10^{-3}
3. 10^7
5. 10^{-2}
7. 4.50×10^2
9. 1.20×10^4
11. 5.26
13. 6.11×10^{-1}
15. 1.47×10^8
17. 4.9×10^{-1}
19. 5.22×10^5
21. 9.99×10^{-2}
23. 3.2×10^5
25. 2.78×10^{-3}
27. 3.17×10^{-1}
29. .00000646
31. 79,800,000
33. 59,900,000,000

Exercise 64

1. $\sqrt{25}$
3. 5
5. $9x$
7. $5y^2$
9. 15
11. $60k^4$
13. 4.243
15. 76.246
17. 21.168
19. $^-1.577$

Exercise 65

1. 3
3. $\sqrt{35}$
5. $4\sqrt{3}$
7. 6
9. $5\sqrt{15}$
11. $^-6\sqrt{21}$
13. $2x$
15. $xy\sqrt{z}$
17. x^6
19. $x^8 y^8$
21. $\dfrac{2x}{y^2}$
23. 108
25. 7
27. $2-\sqrt{3}$
29. 2
31. $\left\{(2, \dfrac{1}{2})\right\}$
33. (a) $b(b-1)$ (b) $c(c+1)(c-1)$
    (c) $(x-6)(x+3)$ (d) $(3x+4)(2x-1)$
    (e) Prime (f) $2x(x^2-2x+25)$

35. $10\sqrt{2}$

37. $28\sqrt{2}$

39. $^-35$

41. $6\sqrt{3}$

43. 270

45. 25

47. 8

49. (a) 8   (b) 36
    (c) $^-4$ (d) $^-1$
    (e) 3   (f) $^-12$
    (g) $^-8$ (h) 8

51. 6

53. $^-x^8$

55. (a) $y = \dfrac{m+2}{a}$ (b) $y = \dfrac{1}{1-x}$

57. $4x^2 + 12x + 9$

59. $14 + 6\sqrt{5}$

61. $^-22$

63. $4x\sqrt{66}$

65. $4a^3 b^3 \sqrt{6ab}$

Exercise 66

1. $\sqrt{2}$

3. $\sqrt{x}$

5. $2x^2$

7. $4\sqrt{15}$

9. $10\sqrt{3}$

11. $6\sqrt{7}$

13. $2\sqrt{3a}$

15. $\dfrac{^-3m}{2}$

17. $\dfrac{4}{5}\sqrt{5}$

19. $\sqrt{3}$

21. $\dfrac{\sqrt{2m}}{4}$

23. $\dfrac{1}{xy}$

25. $\dfrac{1+3x}{1-3x}$

27. $\dfrac{2\sqrt{p}}{p}$

29. $\dfrac{9y^2}{x^4}$

31. $\dfrac{\sqrt{5}}{5}$

33. $\dfrac{\sqrt{26}}{2}$

35. $\sqrt{3} + 3$

41. $\{(7, 8)\}$

43. $\dfrac{\sqrt{2ab}}{b}$

45. $6m^2\sqrt{2m}$

47. $\dfrac{\sqrt{3}}{2}$

49. $2 + \sqrt{x}$

51. $4a^2b^2\sqrt{b}$

Exercise 67

1. $3\sqrt{3}$

3. $3a\sqrt{2}$

5. $\sqrt{2} + \sqrt{3}$

7. 0

9. $7\sqrt{3}$

11. $\dfrac{9\sqrt{2}}{2}$

13. $3\sqrt{6}$

15. $4\sqrt{10}$

17. $3\sqrt{6} + 9\sqrt{3}$

19. $2\sqrt{3} + 3\sqrt{2}$

21. 2

23. $\{^-3\}$

25. $\left\{\dfrac{^-3}{2}, \ ^-1\right\}$

27. $12\sqrt{5}$

29. $\dfrac{6\sqrt{5}}{5}$

31. $\dfrac{31\sqrt{3}}{30}$

33. $\dfrac{9}{4}$

35. $1 - x$

37. $a - b$ means $a + {}^-b$

39. A theorem must be proved whereas an axiom is accepted without proof.

41. Yes.

Exercise 68

1. $\dfrac{\sqrt{6}}{2}$

3. $a\sqrt{x}$

5. $9 - 3\sqrt{6}$

7. $2 + \sqrt{2}$

9. $\dfrac{a\sqrt{b} + b}{b}$

11. $\dfrac{\sqrt{3 - 3x}}{1 - x}$

13. $5\sqrt{6}$

15. $10\sqrt{10}$

17. $\dfrac{a^2 - 2a\sqrt{x} + x}{a^2 - x}$

19. $\dfrac{\sqrt{x^3 + x}}{x}$

21. $\dfrac{\sqrt{4a^4 + a}}{a}$

23. $\dfrac{x^6}{y}$

25. $a^2 + 2ax + x^2$

27. $a^2 + 2a\sqrt{x} + x$

29. $\dfrac{\sqrt{(x^2 - x)(x - 1)}}{x - 1}$

31. $7\sqrt{3}$

33. $^-4\sqrt{5} + 4\sqrt{3}$

35. $a^2 - 2$

37. $\dfrac{x + 1}{x - 1}$

39. $\{^-1, 5\}$

Cumulative Review: Chapters 1 through 10

1.
   $$\begin{array}{ccccc} & \bullet & \mid & \bullet & \bullet \\ & -2 & 0 & 3 & \sqrt{15} \end{array}$$

5. (a) One (b) Two (c) Yes, $x\sqrt{x + y}$

7. $9x^2 - 48x + 64$

9. (a) $\dfrac{2x - 1}{2x - 2}$ (b) $\dfrac{m}{k}$

11. (a) $\dfrac{\sqrt{10}}{2}$ (b) $\dfrac{7}{2x + 1}$

   (c) $a^6 b^9 c^3$ (d) $-9p^2 q + 2pq^2$

13. (a) $a(x^2 + 4)$ (b) $a(x + 2)(x - 2)$
   (c) $a(x + 2)^2$ (d) $a(x^3 - 3x - 70)$

15. 18 inches and 45 inches

17. 18 feet from 40-pound weight

19. $\{1, 2\}$

21. $\left\{\dfrac{180}{7}\right\}$

23. (a) $\dfrac{\sqrt{6} + 1}{5}$ (b) $\dfrac{^-3 + 7\sqrt{3}}{46}$

   (c) $\dfrac{\sqrt{2\sqrt{2}}}{2}$ (d) $\dfrac{x^6}{y^3}$

25. Additive identity

Exercise 69

1. Linear, one variable
3. Quadratic, two variables
5. Linear, two variables
7. 13.0
9. 78.0
11. .550
13. 2.368
15. .0177
17. 5.032
19. 2.689
21. 85.052
23. $\{(5, {}^-8)\}$
25. $\{{}^-3, 7\}$

Exercise 70

1. $\{{}^-3, 3\}$
3. $\{{}^-\sqrt{2}, \sqrt{2}\}$
5. $\{{}^-6, 6\}$
7. $\left\{\dfrac{\pm\sqrt{33}}{11}\right\}$
9. $\left\{\dfrac{\pm 12}{11}\right\}$
11. $\{\pm\sqrt{15}\}$
13. $\left\{\dfrac{\pm 2\sqrt{a}}{a}\right\}$
15. $\{\pm 10\}$
17. $\left\{\dfrac{\pm\sqrt{7a}}{a}\right\}$
19. $\dfrac{1}{2}$
21. $\left\{\dfrac{k^2 + 1}{k - k^2}\right\}$
23. $\{\pm\sqrt{2}\}$
25. $\{{}^-2\}$
27. $\{\pm 6\}$
29. (a) 26.83 (b) $x + 3$ (c) $x^2 + 6x + 9$
(d) ${}^-4x^2 - 8x + 7$ (e) 6 (f) $(a + b) + c = a + (b + c)$
(g) Addition is an undefined term. (h) Look this one up.

Exercise 71

1. $\{{}^-5, 8\}$
3. $\left\{\dfrac{3}{2}, \dfrac{{}^-1}{8}\right\}$
5. $\{0, {}^-1, {}^-2\}$
7. $\{1, 8\}$

9. $\left\{\dfrac{\pm 3}{2}\right\}$ 11. $\{^-6, 2\}$

13. $\{^-1, 7\}$ 15. $\left\{\dfrac{^-3}{8}, \dfrac{1}{2}\right\}$

17. $\{\pm \sqrt{3}\}$ 19. $\{^-1, ^-6\}$

21. $\dfrac{\pm 2\sqrt{a}}{a}$ 23. $\{^-10, ^-1\}$

25. $\left\{\left(\dfrac{17}{3}, \dfrac{8}{3}\right)\right\}$ 27. $x^2 - 3x + 2 = 0$

29. $x^2 + 6x + 8 = 0$ 31. $x^3 - 3x^2 - 4x + 12 = 0$
33. $8x^2 - 6x + 1 = 0$ 35. $9x^2 - 2 = 0$
37. x^{12} 39. $ax^3 + 4ax^2 + 4ax$
41. $\{^-6, ^-3, ^-1, 0, 3 \text{ (Double)}\}$

Exercise 72

1. $\{^-3, 1\}$ 3. $\{^-2, 10\}$

5. $\{^-3, ^-2\}$ 7. $\left\{\dfrac{7 \pm \sqrt{69}}{2}\right\}$

9. $\left\{\dfrac{3 \pm \sqrt{3}}{2}\right\}$ 11. $\left\{\dfrac{3 \pm \sqrt{59}}{10}\right\}$

13. $\{^-3, 3\}$ 15. $\{0, 4, \pm 5\}$

17. $\{4\}$ 19. $\left\{0, \dfrac{5}{2}\right\}$

21. (a) $\dfrac{y}{a}$ (b) $2x^2 + 4x$ 23. (a) $\dfrac{5}{9}$ (b) $2y\sqrt{10x}$

25. $\{6, 12\}$ 27. (a) $\{\pm 1\}$ (b) $\{\pm 3\}$ (c) $\left\{\dfrac{\pm 3}{4}\right\}$

Exercise 73

1. $\{^-2, ^-1\}$ 3. $\{^-10, 5\}$

5. $\{0, 3\}$ 7. $\left\{^-1, \dfrac{1}{4}\right\}$

9. $\left\{\dfrac{^-3 \pm \sqrt{9 - 4a}}{a}\right\}$

11. $\left\{\dfrac{\pm \sqrt{15}}{3}\right\}$

13. $\left\{\dfrac{1 \pm \sqrt{19}}{3}\right\}$

15. $x^2 + 2x - 15 = 0$

17. (a) $\{\pm 2\sqrt{3}\}$ (b) $\{0, 12\}$ (c) $\{\pm 1\}$

19. (a) $^-5\sqrt{5} + 6$ (b) $9 - 18\sqrt{2}$

21. $\{\pm \sqrt{3}\}$

23. $\{0 \text{ (Double)}, 3\}$

25. $\left\{\dfrac{^-1}{4}, 4\right\}$

27. $\left\{\dfrac{^-5}{8}, 1\right\}$

29. 33 dimes

Exercise 74

1. $0 + 8i$

3. $^-5 + 0i$

5. $6 + 4i$

7. $^-\sqrt{2} + i\sqrt{2}$

9. $3 - 2i\sqrt{3}$

11. $5 \pm 4i$

13. $5 \pm \dfrac{i\sqrt{10}}{5}$

15. $2 - 3i\sqrt{2}$

17. $5 + 12i$

19. $^-15 + 20i$

21. $\{\pm \sqrt{3}\}$

23. $\{\pm \sqrt{5}\}$

25. $\left\{0, \dfrac{^-2}{3}\right\}$

27. $\left\{\dfrac{5 \pm \sqrt{34}}{9}\right\}$

Exercise 75

1. $\{\pm i\}$

3. $\{\pm 4i\}$

5. $\{\pm \sqrt{3}\}$

7. $\{^-3 \pm i\}$

9. $\left\{\dfrac{^-1}{2} \pm \dfrac{i\sqrt{11}}{2}\right\}$

11. $\{0, 4\}$

13. $\{(^-1, 3)\}$

15. $\left\{\dfrac{5}{7}\right\}$

17. $\left\{\dfrac{^-4 \pm i\sqrt{14}}{3}\right\}$

19. Yes, if $k < \dfrac{^-1}{p}$

Exercise 76

1. (a) $\sqrt{5}$ (b) 7   (c) $\sqrt{5x^2 + 2x + 1}$
3. 20 miles
5. 48 square feet
7. Yes
9. $^-32$ and $^-48$
11. $42, $54

Exercise 77

1. 8 and 3 or $^-8$ and $^-3$
5. 2 and 12 or $^-6$ and $^-4$

9. Length = 9 feet
   Width = 5 feet
13. 8 nickels

3. 6
7. Length = 10 inches
   Width = 4 inches
11. 9 and 7 or $^-9$ and $^-7$

15. 3016 square feet

Final Cumulative Review I

5. $^-x^2 - 3y^2 + 8z^2$

9. $\left\{ x \le \dfrac{16}{3} \right\}$

13. (a) $\dfrac{5m}{m + 3}$ (b) $\dfrac{m + 3}{m - 5}$

17. (a) $\dfrac{y^2}{z^2}$ (b) $\dfrac{x + y}{xy}$

7. (a) $\left\{ \dfrac{37}{2} \right\}$ (b) $\left\{ \dfrac{^-21}{16} \right\}$

11. $4z(2xy + 5z)(2xy - 5z)$

15. $y = \dfrac{9}{2}$

19. 19.646

Final Cumulative Review II

3. (a) Term   (b) Factor
5. $^-13$

7. (a) $m = \dfrac{2b + y}{3}$ (b) $m = 1 \pm \sqrt{1 + 3a}$

9. (a) $\{x \mid x \leq 5\}$ (b) $\left\{x \mid x < \dfrac{-17}{2}\right\}$

11. (a) $6m(2m + 1)(m - 3)$ (b) $6m^2(4m^2 + 25)$

13. (a) $\dfrac{m - 3}{m^2}$ (b) $\dfrac{3x^2}{x - y}$

15. (a) No (b) $y = \dfrac{-3}{2}$ (c) 6 (d) No. It is not an ordered pair.

17. (a) 4.83×10^2 (b) 6.79×10^3
   (c) 3.61×10^{-3} (d) 1.84×10^{-2}
   (e) 1.5 (f) 1.39×10^{-3}

19. (a) $\{\pm \sqrt{7}\}$ (b) $\{\pm 4\}$

   (c) $\left\{\dfrac{\pm \sqrt{2}}{2}\right\}$ (d) $\{\pm 2i\sqrt{2}\}$

   (e) $\{0, 4\}$ (f) $\{^-3, 0\}$

Final Cumulative Review III

1. $\{3, 4\}, \{3\}, \{4\}, \emptyset$
3. (a) Commutative (multiplication) (b) Distributive
   (c) Associative (multiplication)

5. (a) $\left\{\dfrac{28}{17}\right\}$ (b) $\{\pm \sqrt{3}\}$ (c) $\left\{\dfrac{\pm \sqrt{14}}{2}\right\}$

7. $\left\{x \mid x \leq \dfrac{5}{18}\right\}$

9. $m + 2$

11. $\dfrac{2x^2 - 4x - 6}{x^2 + 2x}$

13. (a) $\{(1, 2)\}$ (b) $\{(2, ^-1)\}$
15. 8 cans of beans, 6 cans of applesauce
17. (a) $4x^2\sqrt{3x}$ (b) $4\sqrt{15}$

   (c) $\dfrac{\sqrt{x + 1}}{x + 1}$ (d) $\sqrt{x^2 + y^2}$

19. (a) $\{\pm i\}$ (b) $\{\pm 1\}$

(c) $\left\{\dfrac{\pm 2\sqrt{10}}{5}\right\}$ (d) $\{0, 4\}$

(e) $\{0, 1\}$ (f) $\{\pm 1\}$

Final Cumulative Review IV

1. (a) 32   (b) 270   (c) 49   (d) 9
3. 8, 10
5. (a) $(x + 10)(x - 1)$ (b) $3(x + 3)^2$
   (c) $(8b - 3)(b - 7)$ (d) $3(x^2 + 25)$

7. (a) $\dfrac{26}{9}$ (b) $\dfrac{2x^2 + x - 1}{2x^2}$

9. 6, 15

11. (a) $\dfrac{3\sqrt{2}}{2}$ (b) $2\sqrt{6}$ (c) $x\sqrt{x}$ (d) $\dfrac{x\sqrt{y}}{y}$

13. (a) $\left\{\dfrac{\pm i\sqrt{2}}{4}\right\}$ (b) $\left\{\dfrac{^-1 \pm i\sqrt{3}}{2}\right\}$

15. Old: Length = 6 feet, width = 3 feet
    New: Length = 17 feet, width = 1 foot
19. (a) Commutative (addition)   (b) Associative (multiplication)
    (c) Identity (addition)   (d) Identity (multiplication)

Final Cumulative Review V

3. (a) $\left\{0, \dfrac{4}{3}\right\}$ (b) $\left\{\dfrac{^-3}{4}, 2\right\}$

5. (a) 9   (b) 8   (c) $^-7$ (d) $\dfrac{1}{32}$

7. Length approximately 18.79 feet, width approximately 5.22 feet
9. 157½ rolls

11. (a) $c = \pm\sqrt{a^2 - b^2}$ (b) $c = \dfrac{\pm\sqrt{a + 2a^2}}{1 + 2a}$

13. 4 and 10

15. (a) $\dfrac{7x + 10}{3x + 5}$ (b) $\dfrac{3 + 2x}{x}$ (c) $\dfrac{^-1}{27}$ (d) $64a^6 x^9$

17. $\dfrac{^-1 \pm \sqrt{1 + kp}}{k}$

19. 7 and 3 or $^-7$ and $^-3$

Index

Abscissa, *126, 127*
Absolute value, *13*
Acute angle, *193*
Addition of equals axiom, *44*
Associative axiom, *7*
Axiom, *2*
   addition of equals, *44*
   associative, *7*
   closure, *6*
   commutative, *7*
   distributive, *8*
   identity, *7*
   inverse, *7*
   multiplication of equals, *44*
   reflexive, *44*
   symmetric, *44*
   transitive, equality, *44*
   transitive, inequality, *63*
   trichotomy, *63*

Binomial, *26*
Brackets, *32*

Cartesian coordinate system, *126*
Closure axiom, *7*
Coefficient, *27*
   numerical, *27*
   literal, *27*

Commutative axiom, *7*
Completing the square, *184*
Complex number, *190ff*
Composite number, *68*
Conjugate, *173*

Degree, *26*
Dependent equations, *137*
Difference of two squares, *74*
Distributive axiom, *8*
Distributive theorem, *31*
Division:
   algorithm for, *37*
   definition, *21*
Double Root, *183*

Equations, *42*
   conditional, *42*
   containing fractions, *113*
   dependent, *137*
   inconsistent, *137*
   linear, one variable, *42*
   linear, two variables, *125*
   linearly independent, *137*
   simultaneous, *137*
Evaluation, *23*
Exponent, *11*
Extreme (of a proportion), *44*

Factor, *11*
Factoring, *68*
Fractions:
    addition, *103*
    building, *101*
    complex, *108*
    division, *95*
    equations containing, *113*
    multiplication, *95*
    reducing, *93*
Fulcrum, *116*

Graph, *127*
    of an inequality in two variables,
      152
    by intercept method, *133*
    of a linear equation, *130*

Half-plane, *151*
Hypotenuse, *193*

"i" form, *190*
Identity axiom, *7*
Imaginary number, *190*
Inconsistent equations, *137*
Independent equations, *137*
Inequality, *62*
Integer, *4*
Intercept, x and y, *133*
Intercept method of graphing, *133*
Inverse axiom, *7*
Irrational number, *4*

Linear, *26*
Lowest common multiple, *100*

Mathematical system, *1*
Mean (of a proportion), *120*
Monomial, *26*
Multiplication of equals axiom, *44*

Negative exponents, *161*

Ordered pairs, *125*
Order of operations, *22*
Ordinate, *126, 127*
Origin, *126*

Parentheses, *32*
Plane, *151*
Plot (a point), *127*
Polynomials, *26*
    addition, *27*
    division, *37*
    multiplication, *35*
    subtraction, *29*
Power, *11*
Prime factor, *68*
Prime number, *68*
Proportion, *120*
Proportional, *120*
    mean, *120*
    third, *120*
Pythagorean theorem, *193, 194*

Quadrant, *128*
Quadratic, *26*
Quadratic equations, *87, 177ff*
    formula, *187*
    pure, *177, 180*
Quadratic trinomial, *81*

Radicals, *165*
    addition, *171*
    division, *168*
    multiplication, *166*
    simplification, *170, 173*
Radicand, *165*
Ratio, *120*
Rational number, *4*
Real numbers, *2*
    addition, *14*
    division, *21*
    multiplication, *19*
    subtraction, *15, 16*
Reciprocal, *7*
Rectangular coordinate system, *126*
Reflexive axiom, *44*

Scientific form, *163*
Scientific notation, *163*
Set, *5*
    empty, *5*
    intersection of, *5*
    null, *5*

solution, *43*
union of, *5*
Set builder notation, *43*
Similar terms, *27*
Simultaneous equations, *137*
Solution set, *43*
Square of a binomial, *75*
Square of a trinomial, *84*
Square root, *165*
   algorithm, *177*
   principal, *165*
   table (Appendix B), *210*
Standard form, *163*
Subset, *5*
Subtraction, *17*
Sum, *11*
Symmetric axiom, *44*
System of equations, *136*
   solution by addition, *140*
   solution by graphing, *136*
   solution by substitution, *143*

Table of squares, square roots, and
     prime factors (Appendix B),
     210
Term, *11*
Theorem, *2*
   on exponents, *158*
   Pythagorean, *193, 194*
Transitive axiom, equality, *44*
Transitive axiom, inequality, *63*
Trichotomy axiom, *63*
Trinomial, *26*

Undefined term, *1*

Variable, definition of, *10*

Zero exponent, *161*